Object Representation and Matching

Based on Skeletons and Curves

DISSERTATION

zur Erlangung des Grades eines Doktors
der Ingenieurwissenschaften

vorgelegt von

Dipl.-Inform. Christian Feinen

geb. am 25.09.1984 in Koblenz

eingereicht bei der Naturwissenschaftlich-Technischen Fakultät

der Universität Siegen

Siegen 2015

Studien zur Mustererkennung

herausgegeben von:

Prof. Dr.-Ing. Heinrich Niemann
Prof. Dr.-Ing. Elmar Nöth

Bibliografische Information der Deutschen Nationalbibliothek

Die Deutsche Nationalbibliothek verzeichnet diese Publikation in der
Deutschen Nationalbibliografie; detaillierte bibliografische Daten sind
im Internet über http://dnb.d-nb.de abrufbar.

ISBN 978-3-8325-4257-3
ISSN 1617-0695

Logos Verlag Berlin GmbH
Comeniushof
Gubener Str. 47
10243 Berlin
Tel.: +49 030 42 85 10 90
Fax: +49 030 42 85 10 92
INTERNET: http://www.logos-verlag.de

Gutachter der Dissertation:

1. Prof. Dr. Marcin Grzegorzek

2. Prof. Dr. Volker Blanz

Tag der mündlichen Prüfung: 10. März 2016.

Acknowledgements

The present work was developed during my time as a member of the Research Group of Pattern Recognition at the University of Siegen. During this time, I was also an associate of the DFG Research Training Group 1564 - "Imaging New Modalities" from October 2010 until June 2014. During these years, I have met plenty of interesting and kind people who contributed to my work either by cooperation, support, meaningful conversation or critical feedback.

I would like to express my gratitude to all of these people. In particular, I would like to thank most sincerely my supervisor *Prof. Dr. Marcin Grzegorzek* for assisting and motivating me throughout the entire period of my studies in his research group. I would also like to thank *Prof. Dr. Volker Blanz* who agreed to be my second supervisor as well as *Prof. Dr. Roman Obermaisser* and *Prof. Dr. Rainer Brück* for being members of commission of my oral exam. Additionally, I am indebted to *Prof. Dr. Andreas Kolb* and *Prof. Dr. Roland Wismüller* who promptly expressed their willingness to act for *Prof. Dr. Blanz* who could not attend the oral examination. A special thanks also goes to *Prof. Dr. Longin J. Latecki* from Temple University. Besides providing me with a great time during an internship in his research group, his work also constituted the foundation for this thesis. Moreover, I would like to mention the *German Research Foundation* for financing my work as a member of the Research Training Group 1564.

Last but not least, I am particularly grateful to all of my colleagues from the Research Training Group 1564 as well as from the Group of Pattern Recognition. I really appreciated working with all of you and I am very happy that we have met each other during my stay in Siegen. A special thanks to those who cooperated with me and/or contributed to the finalisation of this thesis. Finally, I would like to thank my mother *Christina*, my sister *Stephanie* and my girlfriend *Sabine* for supporting and encouraging me during the last years... it helped me a lot!

Abstract

The present thesis is dedicated to the problem of object recognition in the three-dimensional space. This means instead of using exclusively the information that is typically transported by a two-dimensional image, the core concept of this work additionally incorporates the third dimension, namely the depth. This data is captured by a RGB-D or pure depth sensor capable of measuring the distance from the device's position to those objects residing inside its field of view. Almost all projects which are going to be discussed in this work are processing objects in three-dimensional (3D) space. Therefore, it does not matter if these instances were naturally captured or synthetically constructed for the purpose of evaluation.

The actual recognition process has been implemented in analogy to the Path Similarity Skeleton Graph Matching (PSSGM), an object categorisation approach for two-dimensional (2D) objects. The technique taken by itself starts with the skeletonisation of the query and the target to represent these instances by sampling all shortest paths which can be derived from both skeletons, respectively. The notion behind this scanning is the inclusion of geometrical properties of the object's boundary. Finally, the Hungarian method is employed to perform the matching with the aim of calculating the overall similarity between these objects.

The contribution of the current work is now to map the previously described concept into three-dimensional space in order to apply it to 3D objects. Hence, the thesis begins with a deeper investigation of the PSSGM to identify its strengths and weaknesses. In addition to this, a competing set of shape descriptors is introduced to further evaluate the recognition performance of the PSSGM by setting it in relation to this new approach. After encountering skeleton-driven matching techniques as well as skeletal structures in general, two projects are presented able to apply this concept to 3D data. While the first one is only operating on partial views of a certain object (e.g. a chair or a stand), the second method performs the matching on fully segmented vascular structures. Moreover, both approaches substitute the Hungarian algorithm with a more sophisticated procedure covering the problem of finding Maximum Weight Cliques (MWC) inside a graph. The last sub-project of this work addresses the topic of extracting curve skeletons from 3D objects whose geometries rather tend to produce surface skeletons. Besides this, topological features are exploited for the task of object recognition.

In summary, it was possible to successfully map the concept of the PSSGM into the 3D space with the intent to accomplish categorisation tasks. Moreover, several weak points of the original proposal were revealed so that all derivatives were enhanced in terms of accuracy and robustness. Even though all research activities were primarily applied to the development of shape features, capabilities were suggested to enrich the skeleton descriptors for the task of object *instance* recognition. Apart from this, excellent as well as promising recognition results could be achieved in almost all sub-projects during thoroughly planned and executed evaluations. Finally, it has to be noted that the 3D version of the PSSGM has the potential to solve a variety of complex object recognition tasks. However, this proficiency is realised at the expense of a higher processing time and the obligatory use of advanced algorithmics.

Zusammenfassung

Die vorliegende Ausarbeitung widmet sich dem Problem der Objekterkennung im drei-dimensionalen Raum. Im Gegensatz zur ausschließlichen Betrachtung zwei-dimensionaler Bildinformationen verwendet der hier vorgestellte Ansatz die Tiefe als dritte Dimension zur Unterstützung des Wiedererkennungsprozesses. Zur Aufzeichnung dieser Daten kann eine RGB-D bzw. eine reine Tiefen-Messsensorik verwendet werden, die den Abstand, aus-gehend von der Geräteposition hin zu den Objekten innerhalb ihres Erfassungsbereiches, misst. Nahezu alle Projekte, die in dieser Dissertation behandelt werden, verarbeiten Objek-te im 3D Raum. Dabei ist es allerdings unerheblich, ob diese zum Zweck der Evaluierung naturgemäß aufgezeichnet oder künstlich generiert wurden.

Der tatsächliche Prozess der Objekterkennung wurde in Anlehnung an das Path Simi-larity Skeleton Graph Matching (PSSGM) entwickelt, einem Verfahren zur Kategorisierung zweidimensionaler Objekte. Grundlage hierfür ist die Skelettierung des Anfrage- und des Zielobjektes, um diese dann anschließend durch intelligentes Abtasten aller kürzesten Pfade darzustellen. Die Pfade werden jeweils von beiden Skeletten abgeleitet. Der Zweck dieser Technik ist die Einbeziehung geometrischer Eigenschaften hinsichtlich der Kontur beider Objekte. In einem letzten Schritt wird die Ungarische Methode zur Durchführung des Mat-chings eingesetzt und somit die Gesamtähnlichkeit zwischen den Objekten ermittelt.

Der Schwerpunkt der gegenwärtigen Arbeit liegt also in der Überführung des zuvor beschriebenen Konzepts in den 3D Raum, um es dort auf 3D Objekte anwenden zu kön-nen. Hierfür beginnt die Ausarbeitung mit einer genauen Untersuchung zur Identifizierung der Stärken und Schwächen des PSSGM Verfahrens. Darüber hinaus wird ein konkurrie-rendes Set an Deskriptoren vorgestellt, die ebenfalls der Formbeschreibung dienen und zur weiterführenden Beurteilung der Leistungsfähigkeit des PSSGM herangezogen werden. Aufbauend auf der Analyse skelettbasierter Methoden sowie skelettartiger Strukturen im Allgemeinen werden zwei weitere Projekte präsentiert, die das Prinzip des PSSGM im 3D anwenden. Während das erste Projekt ausschließlich auf Teilansichten bestimmter Objekte arbeitet (z.B. Stühle und Tische), erfolgt das Matching der zweiten Methode auf Basis voll-ständig segmentierter vaskulärer Strukturen. Dabei wird zudem die Ungarische Methode in beiden Ansätzen durch eine weitaus anspruchsvollere Prozedur zum Auffinden maximal gewichteter Cliquen innerhalb eines Graphen ersetzt. Abschließend befasst sich die Arbeit mit der Erzeugung von Kurvenskeletten, die für den Einsatz des PSSGM Verfahrens unerlässlich sind. Hierfür werden jedoch eigens 3D Objekte verwendet, deren Geometrieeigenschaften sich eher zur Generierung von Flächenskeletten eignen. Zusätzlich werden topologische Merkmale zur Objekterkennung eingesetzt.

Zusammenfassend ist es möglich gewesen, das Konzept des PSSGM erfolgreich ins Dreidimensionale zu übertragen und dort Kategorisierungsaufgaben durchzuführen. Darüber hinaus wurden mehrere Schwachstellen des ursprünglichen Verfahrens entdeckt, so dass alle weiteren Derivate bezüglich ihrer Genauigkeit und Fehlerrobustheit verbessert werden konnten. Obwohl alle Forschungsaktivitäten primär auf die Entwicklung von Formmerkmalen ausgerichtet waren, wurden Möglichkeiten aufgezeigt, mit denen sich die verwendeten Deskriptoren auch zur Wiedererkennung von Objektinstanzen einsetzen lassen. Zudem konnten vielversprechende Ergebnisse in fast allen Projekten auf Basis sorgfältig geplanter und durchgeführter Tests erzielt werden. Abschließend ist festzuhalten, dass die in dieser Arbeit konzipierte 3D Version des PSSGM das Potential besitzt, eine Vielzahl an komplexen Aufgaben aus dem Bereich der Objekterkennung zu lösen. Hierbei ist allerdings zu berücksichtigen, dass diese Fertigkeiten auf Kosten einer höheren Laufzeit und dem verpflichtenden Einsatz aufwändiger Algorithmik erzielt werden.

Contents

Chapter 1

Introduction

The underlying thesis focuses on the topic of object representation and matching. In more detail, it evaluates graph matching algorithms by representing a wide range of objects based on their skeletons or contour curves. Concurrently with this, the work tries to answer the following questions:

1. "Are (curve) skeletons *generic* enough concerning the task of 3D object *representation?*".

2. "Are (curve) skeletons *meaningful* enough concerning the task of 3D object *matching* and *recognition?*".

3. "What are the *prospects* and *limitations* of such a concept intending to solve the task of 3D object matching?".

Therefore, a wide range of real-world applications are touched encompassing a wide variety of objects in 2D and 3D. These objects are either captured with an depth device or artificially generated as part of a database. In all cases the goal is to *match* or *recognise* a given object (the **query**) to/among a set of other object (the **targets**). These target objects are not necessarily part of the same class, e.g. animals, furniture or tools to mention some of them.

Definition 1.0.1. *Given a set of objects* $\mathbb{D}$, *the query (object)* $q \in \mathbb{D}$ *is that instance which is given to the system in order to place a retrieval request of the most similar candidates in* $\mathbb{D} \setminus \{q\}$.

Definition 1.0.2. *Given a set of objects* $\mathbb{D} \setminus \{q\}$ *and a request in form a query* q, *the target (object) is that instance* p_i *which is currently rated by the system in terms of similarity towards the query.*

The system which undertakes the responsibility to accept the input, to traverse the database, to perform the similarity calculation and to rate it for the purpose of returning a ranked output list, is embodied in an **object retrieval system**. To obtain more information about framework of such a retrieval system, please consult Section 2.2.7.

Object Detection versus Recognition   Frequently, both terms are used synonymously even though their meanings are different. While *object detection* tries to answer the question "**Where** are the objects of interest (or sub-parts of them) in a given (domain-related) scene?", *object recognition* is (mostly) operating on a set of unknown objects responsible for the question "**What** are these objects or to which class/label do they belong?". Thus, object recognition requires a preceding learning phase, e.g. to train the classifier. Another way to describe the relation between these terms is given by Han et al.: "Object recognition differs from object detection in the sense that it does not provide the location of the object but only predicts whether a whole image contains the object or not." [Han+13]

Having this in mind, it is worth mentioning that (multimedia) retrieval systems typically perform recognition tasks. This is also valid for this project. The actual process is realised by performing a matching (cf. Section 2.2.2) that determines the best fitting configuration of correspondences. By taking the costs or rather similarity values of the final alignment, the overall distance or affinity between two objects can be calculated. Contrary to this, Chapter 4 and Chapter 7 are subjected to a slightly different working principle. Here the overall similarity is simply defined by the distance between two feature vectors that have previously been derived from the objects' shapes, respectively. Another exception constitutes Chapter 6, where the matching result is used to register two vessel structures. Instead of determining the similarity between two abdominal aortas, known vessel pairs are aligned towards their 3D pose.

Another aspect worth clarifying concerns the *recognition* term. Although its meaning has already been discriminated from the task of object detection, the objectives of recognition systems can differ between the problem of *object instance* or *object category* recognition. These two classes have to be distinguished in order to receive a better understanding of this work and of pattern or rather object recognition in general.

Object Instance versus Category Recognition   The differentiation between a system performing *object instance* recognition to one that classifies objects with respect to their category is quite simple. The former is designed to identify specific objects (or patterns) which have been trained or seen by the system previously. In contrast to this, the latter recognises the category of an object. Strictly speaking, such a system only returns the class (or category) to which a certain object belongs. This can easily be illustrated by taking a tea cup and a coffee cup as exemplary objects. While the instances of both cups are different, they belong to the abstract class of *cups*. "Category level recognition involves classifying objects as belonging to some category, such as coffee mug or soda can. Instance level recognition is identifying whether an object is physically the same object as one that has previously been seen" [Lai+11b].

Being aware of such a distinction, the reader shall be informed that the present work performs a categorisation of objects. The reason for this is strongly connected to the fact that skeletons are shape descriptors which are natively dedicated to the concept of category-level recognition. Nevertheless, in situations where the object is encompassing a sufficient amount of specific shape characteristics, even a shape-driven feature system is capable of identifying the instance of that object. Please notice that the core idea of the underlying work allows to elegantly assign further unique object properties to the skeletal representation and this, in turn, introduces further capabilities to perform instance classification tasks robustly.

1.1 Motivation

> "When we refer to the 'shape' of an object, we mean those geometrical char-
> acteristics of a specific three-dimensional (3D) object that make it possible to
> perceive the object veridically from many different viewing directions, that is, to
> perceive it as it actually is in the world 'out there'." [Piz08]

Inspired by Zygmunt Pizlo and his notion of shape as well as encouraged by the rapid development of depth or rather RGB-D devices, the idea of this thesis was born. Strictly speaking, the aim was to investigate the recognition performance of 3D objects by extending the perceptual property stimulated by an object's shape with a further dimension, namely the *depth*. This approach is reasonable due to multiple aspects: First, even in presence of *shape constancy* [Piz08], optical properties are potential risks to generate ambiguities which can only be dissolved by changing the perspective on them. Figure 1.1 illustrates some exemplary cases where they occur. Although the objects are still recognisable, such incidents could provoke dangerous situations if autonomously operating machines are involved. Only the incorporation of depth data is able to resolve this ambiguity and to generate view invariant observations. Second, as well as the human eye, visual-based systems suffer from varying light conditions, e.g. shadow edges or front-lighting, whereas depth devices might have a higher insensitivity. Third, the depth data can easily contribute to the process of shape extraction and object clustering. Fourth, having the 3D boundary, new mechanisms can be investigated which might improve the system performance in terms of object recognition and/or representation. Fifth, the alignment of depth with, e.g. RGB data provides a higher amount of information which additionally allows the derivation of correlating characteristics.

Figure 1.1: The figure shows three examples of optical illusions which are only resolvable by changing the perspective[1].

However, being aware of the inherent complexity of a 3D shape, considerations had been made about an adequate concept of handling. At this point the research findings of Bai et al. [BL08] strongly influenced the further process of this thesis. In their work, the authors propose a method capable of reducing the complex 2D shape information by the extraction a one-dimensional (1D) structure, the skeleton that additionally emphasised the most significant parts of the object's shape. Moreover, further boundary information could easily be gathered just by sampling the skeletal branches. By taking into account these advantages together with excellent results which had been obtained by that technique, the decision was made to adapt this concept to the demands of the underlying thesis. Being able to combine the skeletal structure with further unique object properties, e.g. the shape of the 3D boundary, solves the complexity issue and enables new prospects with respect to:

Object Learning & Data Compression   By extracting the curve skeleton of a 3D object, its boundary data is immensely reduced to a composition of multiple connected 1D sequences of 3D points. Each sequence corresponds to a certain skeleton branch that emphasises a shape area significant to the object's appearance. Depending on the working principle of the system, this structure is sufficient to robustly perform 3D object recognition. However, the skeleton can be enriched by further data, e.g. distance to boundary, colour or even multispectral values by attaching it to the corresponding sample locations. A further aspect is the opportunity to generate a *weighted average representation* of an object or even class instead of storing all properties.

Object Matching, Detection & Recognition   Having access to such an object representation, one is able to design smartly and intelligently operating systems. Therefore, well-known (matching) techniques from the field of graph theory as well as its properties

---

[1]http://www.moillusions.com;http://www.illusions.org;http://www.businessinsider.com.au, [online: 19th August 2015]

can easily be employed. It allows e.g. an elegant implementation for the problem of establishing partial matchings in situations where certain object parts are occluded or missing. Moreover, it enables the system to recognise both the instance and the category of an object. As consequence of this, there is no restriction to the task of object detection. Please bear in mind that skeletons naturally count to the class of shape descriptors which are not natively qualified for the job of object instance recognition.

1.2 Contribution

The contribution of this work is dedicated to the matching and (in consequence) to the recognition of 3D objects. Therefore, the advantages and disadvantages of using a skeleton representation coupled with a graph-based matching algorithm are analysed in order to develop accurately and robustly operating systems. The foundation of this work constitutes a popular skeleton-driven matching technique, namely the *PSSGM* approach proposed by Bai et al. in [BL08]. The idea behind this method is quite comprehensible: Using the skeletal structure of an object, the complexity of its shape is reduced by one dimension from 2D to 1D. Moreover, it improves the feature generation process in terms of computation time and discrimination power. The latter is a by-product of the skeletonisation process since the skeletal structure inherently stresses those boundary parts which are mostly contributing to the object's perceptual appearance. In more detail, the actual feature step generation implements the concept of shortest paths which is applied to the skeleton graph. These paths and thus the entire skeletal structure are represented by exploiting an intuitive sampling scheme. Each sample point is then described based on the maximum disk which is fitted into the boundary at this location resulting in a sequence of numbers. Finally, the matching costs are calculated for each pair of skeleton end points in order to determine the optimal alignment between them by invoking the Hungarian method. By taking this knowledge, the underlying thesis presents (separated in self-contained sub-projects) a wide range of applications where the concept of the PSSGM is either re-employed or adapted for the purpose of solving domain related problems in 3D. In the following, the contribution is given in a short summary which encompasses all sub-projects.

Analyses + Databases The starting point of this thesis is made by a thorough analysis of the PSSGM in terms of strengths and weaknesses. A deep inspection of these investigations helped to derive further actions as well as to detect critical issues worth receiving more attention during all subsequent projects. Influenced by this knowledge, further experiments have been made including multiple sets of 2D shapes and databases of 3D objects. This wide variety of instances enabled a comprehensive evaluation of those methods or processing approaches which have been developed in context of this thesis.

Shortest Paths + Skeletons   Coming from the area of 2D shape matching, 3D skeletons are introduced for the task of representing 3D objects. Literature knows two different approaches to extract the skeleton from a 3D object, namely surface skeletons and curve skeletons. This work restricts itself to the latter one. Although the generation of curve skeletons constitutes a more challenging task, its use is reasonable due to the fact that this 1D structure provides less ambiguities and a better substitution for its 2D counterpart. Apart from this, the thesis is also introducing a skeleton related representation based on disconnected 3D curves segments [MYL13]. This special structure emphasised the generic character of the underlying core concept. Moreover, a system guided version of the well-known Dijkstra algorithm has been implemented which accepts a further input parameter describing the general course of an arbitrary path. This adaptation was necessary in order to tackle path ambiguities.

Features + Matching   Although the feature set proposed in [BL08] led to promising results, the application in 3D required more sophisticated descriptors in relation to the specific object domain. Thus, each sub-project introduces its own feature set but keeps the sampling scheme unchanged. Furthermore, topological as well as purely contour-driven descriptors have been considered in contrast to initial ones. Besides these adaptations, further investigations have been made with respect to the matching. Here the originally proposed Hungarian method has been replaced by a novel technique determining Maximum Weight Cliques [ML12]. While the working principle of the MWC approach is more elaborated, its properties and matching results are more attracting than those of the Hungarian method. The thesis shows that this technique provides a powerful and highly flexible instrument for solving challenging matching tasks. At the same time, latent issues of the algorithm are addressed and figured out thoroughly.

Applications   Each sub-project either represents a real-world application or contributes to the improvement of the core concept. Since each application hosts its own problems, the method had to be adapted multiple times to meet the special demands of the domain-related matching tasks. Strictly speaking, the projects are highly heterogeneous and are ranging from 3D chairs to 3D vessel structures. This diversity among these projects demonstrates the generality of the PSSGM. In addition to this, the power of using skeletons or rather boundary-driven shape descriptors could be proved.

1.3   Overview

The underlying thesis is organised in nine chapters. In contrast to other works, this content is separated into sub-projects. However, all projects have in common that they are sharing or contributing to the same core concept which forms the foundation for this thesis. All projects are self-contained and encapsulated in single chapters. Starting with the current one, the introduction, dedicated to the motivation and contribution of the underlying research, the document structure directly hands over to the state-of-the-art section. Besides a holistic view about related publications, Chapter 2 additionally provides an overview of famous depth acquisition techniques, an introduction to a selected set of fundamental side issues and a detailed discussion about skeletons and matching approaches. Altogether, chapter two forms the framework for all subsequent projects which are briefly presented in the following:

Chapter 3   This content part provides a comprehensive introduction to the Path Similarity Skeleton Graph Matching as well as a thorough analysis of its strengths and weaknesses. The outcome of this evaluation is highly important for all subsequent chapters which are using this approach. In particular, the detected weak points received more attention during the development of those recognition tasks falling back to that working principle.

Chapter 4   After encountering skeletal structures, this chapter investigates a further approach for 2D object retrieval. However, instead of using skeletons, *contour points* and *contour curves* are exploited for the task of feature generation. Both primitives are obtained by a novel approach for shape analysis, whereas the feature sets are either adapted or reused. The matching is then realised by the Hungarian method. The objective of this study is to assess the performance of the skeleton towards other shape descriptors.

Chapter 5   This project presents a real-world application focusing on the detection and recognition of 3D object classes, e.g. 3D chairs. Therefore, an arbitrary natural environment is acquired by a depth sensor and all objects inside this scene are transformed into 3D curve segments. Coupled with a sophisticated clustering approach, these curve fragments are considered as a special skeleton type. In consequence of this, the PSSGM can be mapped to this task. By substituting the Hungarian method with a novel matching technique called Maximum Weight Cliques, excellent results have been achieved.

Chapter 6   This chapter is dedicated to the domain of professional medical diagnosis and aims at the registration of two Computed Tomography Angiography series showing the abdominal aorta. While one of them has been captured pre-operative displaying an aortic aneurysm, the second one visualises the result after the surgery. By registering both scans, the physician shall be supported in his or her work of doing the after-care. Therefore, the vascular structures are first extracted by an elaborated segmentation method with the intention of reducing the vascular boundary to its centre line. Finally, these skeletons are matched by exploiting the working principle of the PSSGM together with a new feature set and MWC.

Chapter 7   After receiving a deeper awareness towards the importance of the skeleton quality with respect to the success of the PSSGM. This thesis part is aiming at the extraction of curve skeletons. Therefore, a set of complex 3D objects has been taken into account to evaluate a promising skeletonisation technique presented in [Ren09]. While Chapter 6 facilitates the task of generating curve skeletons due to the almost tubular vessel geometry, the object set which is used in this project is rather qualified to produce surface skeletons. By taking such an opposing set of objects, the performance of the skeletonisation approach shall be stressed. Another characteristic of this chapter is the use of a purely topological oriented feature set which has been employed for the calculation of the objects' similarity.

Chapter 8 finally gathers all results which have been monitored during the evaluation of each single project. The last chapter, Chapter 9, concludes this thesis by summarising all observations which have been collected during the elaboration of this work. Moreover, based on this scientific knowledge, the remaining content discusses further actions that can be derived from it and presents a comprehensive outlook at several future research tasks.

Chapter 2

Related Fundamentals and Approaches

This chapter is dedicated to the broad range of state-of-the-art methods coming from the area of *Object Acquisition*, *Representation* and *Matching*. Being aware of the amount of existing works which have been published over the last decades, only a sliced and narrow band of approaches can be taken into account. Moreover, some of them have been selected to be discussed in more detail since they are closely related to the projects of this thesis. Hence, it is important to gain basic knowledge about these methods to develop a better understanding towards the single topics. In addition to this, a basic introduction is given on how to analyse and how to rate object retrieval systems.

The chapter is structured as follows. Starting with Section 2.1, multiple techniques for acquiring 3D data are outlined. Afterwards, comprehensive knowledge about fundamental concepts and methods, highly related to this thesis, is provided (cf. Section 2.2). State-of-the-art methods will then be discussed in two varying levels of detail. While Section 2.3 is rolling out a holistic overview of the research over the last decades, Section 2.4 is used to present those techniques which are utilised in at least two of the following projects.

2.1 3D Acquisition Techniques

A dominant set of approaches has been proposed based on 2D images captured by Charge-Coupled Devices (CCDs) or Complementary Metal-Oxide Semiconductor (CMOS) chips. In contrast to this, the content below primarily focuses on sensors capable of capturing two-and-a-half-dimensional (2.5D) and three-dimensional (3D) data.

The popularity of depth data is supported by the continuous development of this sensory data in terms of accuracy, housing size and power consumption. At the same time, the corresponding construction costs have been decreased drastically so that the huge consumer market attained access to it. A fact that mainly contributes to the huge success of these sensors. Moreover, different capturing principles have been presented which strongly vary from each other with their own strengths and weaknesses. The following content briefly introduces them and shortly discusses their functional approaches. Therefore, the section is separated into several sub-parts addressing the different concepts for acquiring depth.

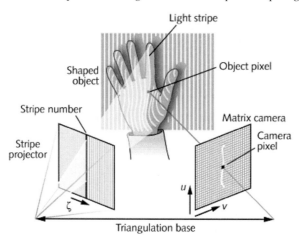

Figure 2.1: The figure shows the measurement principle of the structured light approach using stripes. By capturing the projection of the light pattern, the depth can be recovered by finding the correspondences between the emanated pattern and its projection[1].

2.1.1   Structured Light

The structured light approach is one of the most popular concepts of capturing depth data. It acts in analogy to *stereo vision* systems but incorporates an actively working component. Therefore, the sensor is equipped with two devices, namely a *projector* combined with an ordinary *monochrome camera*. While the projector is emitting a pre-defined light pattern into the world or rather onto the objects of the scene, the camera is capturing it together with the deformations caused by the 3D geometry of these real-world objects (cf. Figure 2.1). This light pattern is typically composed of 2D stripes changing their form and size over time. By

---

[1]http://www.laserfocusworld.com/articles/2011/01/lasers-bring-gesture-recognition-to-the-home.html, [online: 19th August 2015]

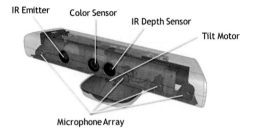

Figure 2.2: The figure provides an overview of the arrangement of the sensors which are part of the Kinect© device.[3]

acquiring this series of different light patterns, the 3D coordinates can easily be recovered by a triangulation approach that detects the displacement of the projected stripes inside the corresponding image. Please bear in mind that such a triangulation approach requires a thoroughly calibrated system setup.

Although the composition of stripes is the most famous pattern in this context, other approaches exist which differ in, e.g. their illumination sources (visible or not visible) or their light patterns (stripes or points). However, the measurement principle behind it remains the same for all of them [SPB04]. Even though the structured light approach is able to produce highly accurate depth images, the devices are not qualified to be used outside. The additional light source and the required sequence of light patterns make the measurement sensitive to sunlight and articulation. Thus, the objects should be static in the sense that they are not allowed to move or to change their form during the acquisition process. Popular market systems are, e.g. the Microsoft Kinect© sensor, the Asus Xtion PRO LIVE© or the 3D DAVID-Laserscanner[2] to mention some of them.

Microsoft Kinect Sensor   The Microsoft Kinect© sensor is one concrete instance of a device exploiting the structured light concept for the purpose of depth reconstruction. Its market launch in 2010 had a profound impact on the area of computer vision regarding both algorithms and applications. Nowadays, it is used to address a broad range of highly heterogeneous issues.

The Kinect© captures RGB images as well as 2.5D data at the same time. Therefore, in addition to an ordinary CCD, the device has a projector capable of emitting near-infrared light, coupled with a near-infrared sensitive camera. All of them are mounted on the same

---

[2]http://www.david-3d.com, [online: 19th August 2015]
[3]https://programmingwithkinect.wordpress.com, [Online: 19th August 2015]

Figure 2.3: The figure illustrates the final depth image and the light pattern used for it. **(Left:)** Depth image recovered by the Kinect© device. **(Right:)** The dot pattern utilised for the purpose of triangulation. Moreover, one can clearly recognise those object's regions which are absorbing the light due to their physical properties.

device. The latter enables the acquisition of depth data based on triangulation using the structured light approach. Moreover, a microphone is included to enrich the information with an audio stream. However, acoustic signals do not play any role in this work. Figure 2.2 provides an overview of the hardware's layout. In contrast to the classical stripes, the Kinect© utilises a pseudo-random dot pattern as shown in Figure 2.3. By capturing the projection of that structure, the 3D coordinate is recovered by identifying dot correspondences based on the original and undisturbed pattern.

2.1.2   Time-Of-Flight

Time-of-Flight (TOF) is another concept aiming at the depth estimation process. In analogy to the structured light measurement, Time-of-Flight devices are also equipped with a light source and a camera. In contrast to the structured light approach, pulses of laser light or a continuously modulated light in terms of frequency or amplitude is emitted (cf. Figure 2.4). By monitoring this light, the time of travelling or the phase shift can be measured indicating the way from the emitter to the object and back to the camera. Compared to the techniques using the concept of structured light, these sensors suffer from a lower resolution and the problem of so-called *Flying Pixels*.

Flying Pixels   These visual 3D scene artefacts mostly occur at the edges of the object's geometry. At these locations, the depth information is oscillating around the distance

of the object (foreground) and the distance of the background. Both the identification
and the correction are difficult tasks and require highly sophisticated strategies. Some of
these techniques are addressed in [SK10] discussing their strengths and weaknesses in detail.

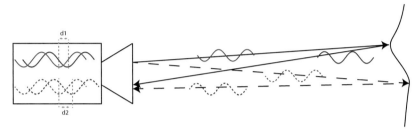

Figure 2.4: The figure illustrates the measurement principle behind the modulated light
concept. Ranges with a higher distance lead to a larger shifts in their phases compared to
ranges with lower distances. However, due to the symmetry property of sinusoidal wave
form, the signal becomes ambiguity if the shift is greater or equal to 2π.

In addition to this, multi-path reflections or sunlight are potential candidates to affect
the measurement in form of interferences (depth errors or noise). Hence, the TOF technique
does not provide the same accuracy as the structured light-driven approaches. Nevertheless,
the actual acquisition process is highly suitable for applications with higher dynamics due
to its capturing speed, e.g. in the automotive area.

2.1.3   Stereo Vision / Laser Range Measurement

These two measurement concepts are highly similar to the ones mentioned above. While
the *stereo vision* approach works in analogy to the structured light technique, the working
principle behind a *laser rangefinder* is similar to that of a TOF device. Please further notice
that these methods even constitute the foundation for the structured light and time-of-flight
sensor architecture and not the other way around.

Stereo Vision Systems   Instead of using a camera-projector pair, stereo vision systems
are operating on two cameras arranged in opposite directions but with known distance
to each other (called *baseline*). Moreover, this technique ranks among the class of passive
(optical) measurement principles based on light reflection. To obtain a 3D world coordinate,
correspondences have to be established between the pixels in the left image with those in

the right one or vice versa. Figure 2.5 shows the setup of such a system. In order to find these matchings, a mathematical model is used which is well known as *epipolar geometry*. Using this model, computational efforts can be reduced drastically while improving the accuracy of the measurement. Strictly speaking, the detection of correspondences is relaxed to a single line in one of the images. Afterwards the offset, or rather *disparity*, between all pixel pairs is estimated. In combination with baseline as well as other extrinsic and intrinsic parameters, all magnitudes are available to perform the depth recovery by triangulation. For further details, the reader is referred to [HZ03; TV98] providing an in-deep introduction to this topic.

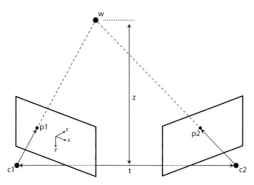

Figure 2.5: The figure illustrates the setup of a stereo vision system. On the left and on the right the image planes are shown. A certain 3D world coordinate w is then estimated by triangulation. Therefore, the two 2D projections of w, namely $p1$ and $p2$, have to be localised in both images. Having $p1$ and $p2$, their disparity can easily be derived and the depth is calculated by means of additional extrinsic and intrinsic parameters.

Laser Range Measurement   A laser range-finder can been viewed as a dimension reduced TOF device recovering the depth by emitting a pulsed laser light or a constant modulated laser beam. The round trip of this light, from the source to the object and back to the detector, is then measured based on the time or the phase shift of this signal. For example, if the speed of light is assumed to be $a = 3*10^8\,m/s$, then one nanosecond would correspond to $0.3\,m$ which has to be measured by the receiver unit (light pulse). The actual distance is obtained by the formula: $(a \cdot t)/2$. A broad overview of existing measurement principles and their applications is given in [PB07].

Laser range-finders are active (optical) measurement devices determining the depth by the reflection of the previously emitted light. The sensor's head typically consists of a single line or array detector coupled with two mirrors rotating in azimuth and elevation. On the one side, this construction increases the field of view of the device. On the other side, it also enables the generation of 3D point clouds. Depending on the desired accuracy, a measurement might encompass thousands of 3D world coordinates.

2.1.4 X-Ray Computed Tomography and Magnetic Resonance Imaging

Coming from the broad consumer market, this section is devoted to a more specific application area in computer vision, namely *medical imaging*. Related to this work, two *computer-aided diagnosis* systems are presented, the Computed Tomography (CT) and the Magnetic Resonance Imaging (MRI). Although both acquisition techniques are able to deliver depth information like the approaches above, their working principles are entirely different (even to each other). While the previous methods exploit visible or non-visible light to measure, e.g. the reflection at a certain surface point of an arbitrary object, the CT and the MRI device is focusing on visualising the interior of a non-transparent object, typically the human body. Therefore, the CT system rotates a receiver and an x-ray emitter around the object of interest to obtain depth data from multiple perspectives. In contrast to this, the MRI is using strong magnetic fields together with radio waves to activate magnetic properties of hydrogen nuclei in order to align them. Additionally to the visualisation of bones, organs and other tissue, the output data can be employed for the purpose of recovering 3D models.

Besides a controversial debate regarding health hazards in respect to the acquisition process, both sensors have different strengths and weaknesses. CT systems, e.g. are good for analysing bony structures, lungs, cancer and pneumonia. The MRI has better properties in displaying soft tissue like ligaments, tendons and the brain. Moreover, it is worth knowing that CT systems are more frequently used than the MRI due to lower costs and a better acquisition performance.

Computed Tomography   Computer imaging system based on the classical radiography. Instead of capturing only a single image, the CT generates multiple image slices of the body or rather the region of interest (ROI) by rotating the x-ray source and its receiver around that ROI (cf. Figure 2.6). Like the classical radiography, these images are obtained by overlaying consecutive structures. Moreover, both disciplines exploit the fact that the absorption of x-ray photons differs between the varying organs and tissue inside the human body. Thus, the actual acquisition process does only measure the intensity of transmitted photons after passing various tissue and bones leading to darker or brighter areas in the x-ray image or slice. The degree of absorption is estimated by an inverse Radon Transform [Bal12]. For this purpose, so-called absorption profiles are generated for each position that is reached by rotating the x-ray source. The final images can then be viewed separately to support the process of diagnosis. Moreover, and even more interesting for this work, they are also utilised for reconstructing a 3D model. It is worth knowing that both the size and the resolution of this model are configurable.

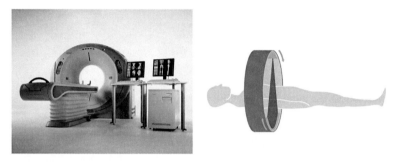

Figure 2.6: The figure is dedicated to the Computed Tomography system. **(Left:)** A picture of such a device showing the most famous layout of it. **(Right:)** Abstracted illustration of the working principle behind a CT scanner where the emitter and the receiver can be rotated around the region of interest in order to produce multiple image slices.[4]

Magnetic Resonance Imaging In analogy to a CT, the MRI technique follows the same objective of visualising the interior of the human body to physicians (and researchers). In contrast to the CT, this technique can only be applied to living beings caused by the fact that the method strongly relies on the magnetic properties of hydrogen nuclei. Strictly speaking, instead of using x-radiation like a CT, MRI scanners exploit powerful magnetic fields combined with radio frequencies to align these nucleus dynamically. Without going to deep in detail, MRI images do not display the degree of absorption caused by bones and tissue. Encouraged by the radio input waves, the hydrogen atoms start to emit radio waves by themselves. These waves are then captured for the purpose of generating images.

"The hydrogen nuclei behave like compass needles that are partially aligned by a strong magnetic field in the scanner. The nuclei can be rotated using radio waves, and they subsequently oscillate in the magnetic field while returning to

---

[4]http://www.radiologie-sanderbusch.de; http://www.medicalradiation.com, [online: 19th August 2015]

equilibrium. Simultaneously they emit a radio signal. This is detected using antennas (coils) and can be used for making detailed images of body tissues."[5]

This output in form of radio waves is controlled by the power of the electromagnetic fields as well as by the radio frequency that is externally introduced to it. Moreover, the contrast of the image is strongly influenced by the power of the frequency and the magnetic fields. Thus, different types of tissue are emphasised just by adjusting these energies carefully.

2.2 Fundamentals

All projects being part of this work are utilising methods or trying to approach problems which are well known in the field of computer science. However, many are accommodated in other research fields, e.g. *graph theory*. Being aware of this, the following content provides a rough framework for all subsequent topics. It does not claim to present a complete overview of these fundamental subjects since this would go beyond the scope of this study. Please keep in mind that only a selected set of information is provided to support a better understanding towards the projects which will be presented in the following chapters.

2.2.1 Basic Introduction to the Assignment Problem

The *assignment problem* is the most famous one in context of object recognition. Its native formulation is highly generic and can be transferred to many other (real-world) applications. The most popular work concerning the assignment problem has been proposed by Kuhn in 1955. His article [Kuh55] with the title "The Hungarian Method for the assignment problem", proposes a solution to this concrete instance of combinatorial optimisation problems and still belongs to the state-of-the-art in nowadays research. Kuhn describes the assignment problem as follows:

> Given N agents and an equivalent number of tasks, where each agent $p_{i=1,...,N}$ can be assigned to any task $q_{j=1,...,N}$. Moreover, let $\mathbf{C}$ be a predefined cost matrix carrying the costs of each assignment $x_{i,j}$. The goal is now to find an optimal alignment with minimum total cost under the constraint that exactly one agent performs exactly one job. (cf. [Kuh55])

This scenario is better known as the *classical assignment problem*, which can mathematically be expressed as [Pen07]:

---

[5]http://www.drcmr.dk, [online: 19th August 2015]

$$\text{argmin} \sum_{i=1}^{N} \sum_{j=1}^{N} C_{i,j} x_{i,j} \quad , \tag{2.1}$$

subject to:

$$\sum_{i=1}^{N} x_{i,j} = 1 \quad j = 1, \dots, N \quad , \tag{2.2}$$

$$\sum_{j=1}^{N} x_{i,j} = 1 \quad i = 1, \dots, N \quad , \tag{2.3}$$

where $x_{i,j} \in \{0,1\}$ with $x_{i,j} = 1$ if the agent p_i corresponds to the task q_j and otherwise zero. Strictly speaking, this assignment is a bijective mapping operating on a bipartite graph (cf. Section 2.2.2 and Section 2.4.2.1). It can be considered as a special case of the *minimum cost flow problem* and this, in turn, forms an instance of a linear program, which will be discussed in Section 2.2.4.

Besides this, further derivations of the assignment problems have been developed over the last decades, e.g. the quadratic assignment problem (QAP) having a quadratic objective function [ZS14; Bur+98]. In contrast to this, the linear bottleneck assignment problem (LBAP) minimises the maximum among all assignments and the generalised bottleneck assignment problem (GAP), the most abstracted version of the linear assignment problem (LAP), allows the establishment of one-to-many correspondences. A broad overview with further information is given in [Pen07].

2.2.2   Basic Introduction to Graph Theory

The graph theory is strongly connected to the previously introduced assignment problem. Moreover, it can be exploited for the purpose of determining the shortest paths. This section delivers a fundamental introduction towards the terms and definitions required to support a better understanding regarding the principles of the graph matching algorithms employed in this work. Please notice that all explanations are primarily following the content of "Graphentheoretische Konzepte und Algorithmen" [KN09] and "Algorithmische Graphentheorie" [Tur09]. Thus, the interested reader is referred to these books in order to retrieve further information.

Definition 2.2.1. *A graph, $G = (A,B)$, is an abstract structure defined as composition of vertices $v_i \in A$ and edges $e_j \in B$. A is a non-empty, (not necessarily finite) set of graph vertices, which are also* *known as nodes. These nodes are partially or fully connected by the set of edges B (sometimes called links). Thus, an edge represents a binary relation of an ordered or unordered pair of vertices (v, v').*

Hence, a graph can be viewed as a network that constitutes a powerful instrument for modelling connection-based concepts whose objects or other entities are represented abstractly by the graph's vertices. In addition to this, a weight can be assigned to each edge that indicates an arbitrary significance of this connection, e.g. a cost or distance value. The edges are either *directed* or *undirected*. This means the edge in a directed graph has a clear direction defined by exactly one predecessor and one successor. Consequently, an ordered relation is not symmetric. In contrast to this, the links inside an undirected graph are without orientation and thus, there is no order among the pairs of nodes: $v, v' \in A \rightarrow (v, v') = (v', v)$.

Moreover, graphs can be visualised perfectly in the Euclidean space. A simple illustration of such a diagram only encompasses lines/arrows and circles. Although there might be multiple variations concerning the arrangement of these entities, the diagram defines the underlying graph uniquely. However, the shape or the length of a single element is meaningless towards its interpretation.

Adjacency Two vertices are adjacent to each other if they are connected by the same edge. This special kind of relation is used for modelling conceptual correlations of the underlying scenario. For example, the shortest path can only be established between nodes which are adjacent to each other. In this context, vertex-edge or edge-edge pairs are incident if the edge connects to the node or if two edges are sharing the same vertex.

In context of this thesis, graphs are primarily employed for solving the assignment problem or determining the shortest paths. Even though the problems are different, their solutions are respectively formed by a subset of edges: $B \subseteq B$. This subset is typically called *matching* for a certain solution of an arbitrary assignment problem.

Definition 2.2.2. *A matching $\mathcal{M}$ is a subset of graph edges. An edge fulfils the condition of being a valid correspondence if there is no other edge incident to it. Strictly speaking, each vertex is only connected to one edge.*

Matchings can be classified according to the following types: *maximal, maximum* and *complete* (cf. [Tur09]). A maximal matching cannot be further extended without violating the matching definition above. However, this type is not necessarily a maximum one, whereas maximum matchings are always maximal due to their definition of being the largest sets of edges. Finally, a complete correspondence configuration is characterised by the property that all vertices are encompassed by the matching. Thus, it is maximum and maximal. Like other works, this thesis does also use the term *partial matching* to refer to a correspondence set that excludes graph nodes from the final result.

Please notice that the matching term is used ambiguously. On the one hand, it points to the subset of graph edges and on the other to the actual *task*. A concrete example is the expression *shape* or *object matching*, namely the problem of determining the (dis-)similarity between two shapes or objects. Moreover, it is worth knowing that the classical assignment problem is typically solved based on a bipartite graph.

Definition 2.2.3. *A bipartite graph is a k-partite graph with $k = 2$.*

Definition 2.2.4. *A k-partite graph is characterised by two properties. First, its vertices can be partitioned into k disjoint subsets $A_1, \ldots, A_k$. Second, its nodes are only adjacent to each other if they are residing in different subsets: If $v \in A_i$ and $v' \in A_j$ and v adjacent to v' then $i \neq j$.* (cf. [Tur09])

2.2.3   Basic Introduction to Dynamic Programming

Dynamic programming (DP) is a powerful tool for solving algorithmically complex optimisation tasks by breaking them down into smaller but homogeneous sub-problems. This fragmentation leads to a relaxation of the original problem due to the fact that these sub-problems are easier to solve than the original one. The overall solution of the given problem is then obtained by combining all partial solutions from the previous steps. This relation is known as *Bellman's Principle of Optimality* and must be ascribed to the mathematician Richard Bellman who introduced the theory of DP [Bel54].

Please notice that the term DP has an inherent ambiguity. On the one side, it refers to the mathematical optimisation process and on the other side to the act of computer programming. However, independent from its meaning, DP is always dedicated to the task of simplifying a given problem by decomposing it into smaller sub-parts. Although this procedure sounds similar to *recursion*, DP does not perform any recalculation of sub-problems which have already been touched in previous cycles. Instead of this, all results are going to be stored and reused in later stages (*overlapping sub-problems*). Thus, the DP principle highly improves the computational speed compared to the native concept behind recursion. However, it has to be stressed that every optimisation problem has to fulfil at least two conditions in order to be solved by means of the DP, namely the property of having an *optimal substructure* and *overlapping sub-problems*. Famous instances are, e.g. the problem of finding the shortest path between two nodes inside a directed acyclic graph (DAG) or the calculation of the Fibonacci sequence.

Literature suggests two different ways for the purpose of partitioning and solving a given problem, namely *top down* and *bottom up*. While the *bottom up* approach first analyses it in order to detect less complex sub-parts, the *top down* operation works in analogy to a lazy evaluation paradigm. Thus, the given problem is broken down from the top and the method processes the sub-parts sequentially in the order in which they occur. Consequently, the *top down* technique does not necessarily start with the easiest one in this hierarchy like the *bottom up* principle. Indeed, this behaviour is more similar to the native recursion but in contrast to it, all calculations are stored with the intention of looking them up if they are re-discovered. Hence, the *top down* approach is also known as *memorisation*. The Dynamic Time Warping (DTW) (Section 2.4.3.1) is one representative of a DP-driven algorithm.

2.2.4 Basic Introduction to Linear Programming

Linear programming is a popular optimisation technique to maximise or minimise a certain linear, real-valued objective function over a domain. This domain, in turn, is defined by a couple of linear constraints which are modelled in form of linear equalities or inequalities.

> "The objective function is linear, and the domain, or feasible set, is defined by linear constraints."[6]

Thus, linear programming is also known as *linear optimisation*. Even though the term *linear* is well-known, it can be expressed as follows[6]:

$$g(a\mathbf{x}_1 + b\mathbf{x}_2) = a\, g(\mathbf{x}_1) + b\, g(\mathbf{x}_2) \quad , \tag{2.4}$$

where a and b represent two constant values and $\mathbf{x}_1$, $\mathbf{x}_2$ are two vectors. Assuming all variables are not negative and the linear constraints are expressed as inequalities, the standard maximum linear program can be formulated as[7]:

$$\operatorname{argmax} \quad \mathbf{a}^\mathsf{T}\mathbf{x} \quad , \quad \text{subjects to} \tag{2.5}$$

$$\mathbf{A}\mathbf{x} \le \mathbf{b} \quad , \tag{2.6}$$

$$\mathbf{x} \ge 0 \quad , \tag{2.7}$$

where Equation (2.5) refers to the objective function and $\mathbf{A}$, $\mathbf{b}$ to the feasible set. While $\mathbf{a}$ depicts a vector of known constants, the values of $\mathbf{x}$ have to be determined during the maximisation process.

Please keep in mind that a linear program might be more heterogeneous, e.g. by encompassing equalities *and* inequalities. However, based on the standard maximum problem, the

---

[6]http://www.cs.columbia.edu/coms6998-3/lpprimer.pdf, [online: 19th August 2015]
[7]http://www.math.ucla.edu/~tom/LP.pdf, [online: 19th August 2015]

actual working principle can easily be mapped to a (convex) n-polytope (where n determines the dimension of the embedding space). This polytope represents the region accommodating the point that maximises the objective function. Moreover, the edges of the polytope are specified by the intersections of the half spaces defined by the inequality constraints (assuming that the feasible set is bounded). The maximum or minimum value of the objective function is then supplied by one of the corner points of this polytope (or feasible region).

In addition to this, every linear program (LP) (in the following named *primal*) always originates another linear program associated to it which is known as the *dual* or *dual problem* of the primal. This dual linear program can easily be retrieved by executing a number of pre-defined transformation steps[8]. The dual to the primal introduced in Equation (2.5), (2.6) and (2.7) is defined as [9]:

$$\text{argmin} \quad \mathbf{b}^\mathsf{T}\mathbf{y} \quad , \quad \text{subjects to} \tag{2.8}$$

$$\mathbf{A}^\mathsf{T}\mathbf{y} \geq \mathbf{a} \quad , \tag{2.9}$$

$$\mathbf{y} \geq \mathbf{0} \quad . \tag{2.10}$$

"The optimum of the dual is now an upper bound to the optimum of the primal."[9]

This duality constitutes the foundation for multiple algorithms capable of solving LPs, e.g. the *branch and cut* [Mit02]. Nevertheless, the most famous technique for processing an linear program is the so-called *simplex method* [DT97]. It is worth knowing that the simplex algorithm does solve the dual problem in addition to the given one. Compared to *quadratic programs (QPs)* (cf. Section 2.4.2.2), LPs are of simple nature in terms of *computational complexity* and can be solved in polynomial time.

2.2.5   Basic Introduction to the Shortest Path Problem

Although the term is somehow self-explaining, this section shall be used to provide a more concrete explanation of the underlying problem and its applications. Moreover, the working principle of Dijkstra algorithm is roughly outlined as one representative for calculating the shortest paths.

Definition 2.2.5. *Given a weighted graph G and the two vertices $a, b \in G$. Then the shortest path problem (SPP) describes the task to determine a path ρ between a and b with the minimum weight sum. (cf. [Tur09])*

---

[8]http://www.cs.columbia.edu/coms6998-3/lpprimer.pdf, [online: 19th August 2015]
[9]http://theory.stanford.edu/~trevisan/cs261/lecture06.pdf, [online: 19th August 2015]

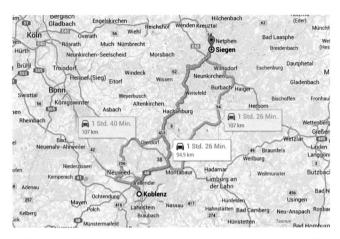

Figure 2.7: The figure illustrates a concrete instance of the shortest path problem. In this application, the optimal way between two cities which can be travelled by car is searched[10].

In more detail, a shortest path ρ can be viewed as a sequence of edges $\{e_1, e_2, \ldots, e_N\}$ in G, which are pairwise incident to each other. This set then connects a with b while minimising the sum of edge weights:

$$\rho = \underset{\hat{\rho} \in \mathcal{K}}{\arg\min} \sum_{i=0}^{N} \psi(e_i) \quad , \quad (2.11)$$

where $\mathcal{K}$ depicts the set of all valid paths $(\hat{\rho}_j)$ connecting $a, b \in G$ and $\psi(\cdot)$ returns the weight of the edge e_i. In literature, this is called the *single-pair* shortest path problem. Besides that, there also exist the *single source* and the *all-pairs* SPP[11]. A more specialised instance is given by the *shortest path tour problem*. It is defined on a directed graph with non-negative arc lengths and its path has to cross a given sequence of node subsets [Fes09]. Generally, there is no restriction on the edges and thus, they can be directed or undirected. The weights are commonly expressed by real-valued numbers which might also be negative. The meaning of these weights strongly depends on the underlying application and ranges from *costs* over *time* to *distances* and *probabilities* (cf. [Tur09]).

---

[10]https://www.google.de/maps, [online: 19th August 2015]
[11]http://www.cs.cmu.edu/~15451/lectures/lec9.pdf, [online: 19th August 2015]

Geodesic   The geodesic shortest path problem (GSSP) is highly related to the ordinary one. A *geodesic* indicates the shortest path between two points on a surface and should be viewed as a length minimising curve. However, the determination of such geodesic is mostly realised by establishing a graph structure. In this thesis, both terms are used interchangeably.

Even though it might sound trivial to solve an instance of the shortest path problem, it manifests a complex task requiring sophisticated strategies. While Figure 2.8 illustrates an arbitrary graph and its shortest path, Figure 2.7 provides a more practical example of the SPP. Please notice that this section is only roughly touching this topic, it would go far beyond the scope of the thesis to provide a comprehensive discussion. In order to obtain more information, the work of Kairanbay Magzhan and Hajar Mat Jani [MJ13] shall be mentioned here. Their work provides a review and an evaluation of shortest path algorithms encompassing the Dijkstra, the Floyd-Warshall, the Bellman-Ford and the Genetic algorithm.

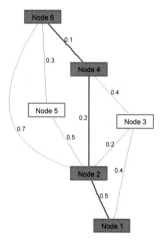

Figure 2.8: Shortest path example based on a fictive graph and constructed edge weights. The red path indicates the shortest path between node 1 and 6. By summing up all associated edge weights, the shortest path leads to minimum costs compared to the other ones.

Dijkstra Algorithm   The technique was first published in the year 1959 by Edsger Dijkstra [Dij59] and counts to the most popular algorithms to calculate shortest paths. It is also employed in this thesis and thus, the technique shall briefly be outlined in the following. Its basic working principle is composed by a mixture of both a *greedy* and a *dynamic programming* approach. By selecting adjacent graph nodes smartly during the computation of the optimal path, the algorithm traverses each vertex only once. The actual implementation is straightforward and can easily be explained. Therefore, let G be a directed graph with non-negative edge weights $\psi(e) \geq 0$. In each cycle, one graph node is processed and never revisited again. Starting at $a \in G$, the algorithm first calculates the distance to all of its adjacent nodes and marks a as "visited". Subsequently, the vertex with minimum distance is selected and again the algorithm computes the distance to all of its neighbours. Therefore, it reuses the distance information of those paths which have been already calculated in previous steps. With other words, assume that v has the shortest distance to the start point (a) and that $d(a,v)$ returns the total length to it. Moreover, let $\hat{v}$ be an unvisited, adjacent node of v whose connection is established by the edge e_j. Then the distance of $\hat{v}$ is retrieved by:

$$d(a,\hat{v}) = d(a,v) + \psi(e_j) \quad . \tag{2.12}$$

If $\hat{v}$ has already been labelled with a distance value from another node, the minimum of both is taken. Once all adjacent nodes have been traversed, the algorithm labels v as "processed" and proceeds by selecting the node with minimum distance to the start vertex. This procedure is repeated until the Dijkstra method reaches its destiny node $b \in G$. Figure 2.9 demonstrates the first two steps of the algorithm operating on a constructed graph structure. The actual shortest path is finally determined based on a backtracking procedure.

2.2.6 Basic Introduction to the Distance Transform

This section briefly covers the topic of the *distance transform (DT)*. Fundamental knowledge towards this operation is helpful since it is frequently employed in this work. Section 2.4.1, e.g. discusses the use of the distance transform in the context of skeletonisation while Chapter 7 extends the DT to the so-called *(extended) feature transform*.

> "The central problem of a distance transform is to compute the distance of each point [inside the grid] to a given subset of it." [Fab+08b]

Thus, the basic task of a distance transform is quite simple: Given a grid-based input structure $\mathcal{D}$, e.g. a digital binary image, whose grid points $\mathbf{p}_i \in \mathcal{D}$ either are labelled as region of interest $\mathcal{D}^+ \subset \mathcal{D}$ or not $\mathcal{D}^- = \mathcal{D} \setminus \{\mathcal{D}^+\}$. The DT determines the smallest distance of each grid point $\mathbf{p} \in \mathcal{D}^-$ to its closest neighbour $\mathbf{q} \in \mathcal{D}^+$:

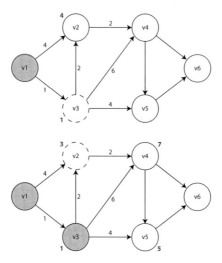

Figure 2.9: The figure shows the results after the first and the second iteration of the Dijkstra algorithm. **(Top:)** Starting with v1, the edge distances are propagated to v2 and v3. Afterwards, v3 (dashed) is selected due to its minimum distance to the start node. **(Bottom:)** Like in the previous step, the distance information is pushed to the neighbours of v3. Please observe that the old distance value (4) of v2 is overwritten in order to keep the shortest distance to v1.

$$D(\mathbf{p}) = \operatorname*{argmin}_{\mathbf{q} \in \mathcal{D}^+} d(\mathbf{p}, \mathbf{q}) \quad , \tag{2.13}$$

with $d(\cdot, \cdot)$ being an arbitrary distance function. Moreover, the result of the DT has the same structure as its input but differs in the meaning towards the single values of each grid point. Figure 2.10 shows a numerical example of the distance transform by incorporating the Euclidean distance. Please notice that the values have been squared for the purpose of better readability. Caused by its visual representation in 2D, the distance transform is also known as *distance map* or *distance field*.

It is worth knowing that the distance values can also be signed. The signed distance map is of special interest whenever a point has to be classified according to its position with respect to the region of interest. Strictly speaking, the value is negative if the point resides outside the ROI and positive if not (or vice versa). Such a characteristic is required in context

0	0	0	0	0	0	0	0	0
0	0	0	0	0	0	0	0	0
0	0	0	1	1	1	0	0	0
0	0	1	1	1	1	1	0	0
0	0	1	1	1	1	1	0	0
0	0	1	1	1	1	1	0	0
0	0	0	1	1	1	0	0	0
0	0	0	0	0	0	0	1	0
0	0	0	0	0	0	1	0	0

0	0	0	0	0	0	0	0	0
0	0	0	0	0	0	0	0	0
0	0	0	1	1	1	0	0	0
0	0	1	2	4	2	1	0	0
0	0	1	4	8	4	1	0	0
0	0	1	2	4	2	1	0	0
0	0	0	1	1	1	0	0	0
0	0	0	0	0	0	0	1	0
0	0	0	0	0	0	1	0	0

Figure 2.10: Numerical example of the distance transform by incorporating the Euclidean distance. (Left:) A binary input image indicating the region of interest by the value 1.0. (Right:) The resulting distance map is carrying the information as squared Euclidean distance values. (cf. [Fab+08b])

of shape representation, where the region of interest is specified by the shape's boundary (cf. [Fri+00]). However, the calculation of a distance map usually involves more sophisticated techniques. A comparative survey of 2D Euclidean distance transform algorithms is given in [Fab+08b], whereas [Bor96] and [HF07a] are covering the computation in 3D. In the following, the method proposed in [HF07a] shall be used as a concrete example for all subsequent explanations. Here the authors introduce an approach for solving the signed distance function (depicted as $f(\cdot)$) by taking into account its implicit gradient definition:

$$\|\nabla f(\mathbf{x})\| = 1 \quad . \tag{2.14}$$

Equation (2.14) asserts that the length of the gradient equals one at any point where the function is differentiable. This is also known as *Eikonal equation* or *Hamilton-Jacobi equation* (on Cartesian domains). Using this property, the calculation of the distance map can be implemented by tracking a moving interface in a multi-dimensional space, so that *distance = speed × time* (cf. [HF07a]):

$$\|\nabla T\| v = 1 \quad , \tag{2.15}$$

with $v = \frac{dx}{dT}$ being the speed that only depends on the position $\mathbf{x}$. Furthermore, $T(\mathbf{x})$ determines the arrival time of the interface when it crosses the point $\mathbf{x}$. The most famous representatives using this wave propagation model are *Fast Marching Methods (FMMs)* [Set95] and *Level Set Approaches* [OS88]. Having this context, the work of [HF07a] has to be viewed as an improved version of the FMM with the name *multistencils fast marching (MSFM)*.

> "[The MSFM] computes the solution at each grid point by solving the Eikonal equation along several stencils that cover its entire neighbour points and then picks the solution that satisfies the upwind condition." [HF07a]

Fast marching as well as level set techniques are also subject of a deeper discussion in Chapter 6. Please consult the corresponding sections to obtain more detailed information about these topics.

2.2.7 Basic Introduction to the Rating of Retrieval Systems

This work aims at the development of matching algorithms. These techniques, in turn, are operating on (dis-)similarity measures. Some of them, e.g. the Euclidean distance, are trivial for a single pair of elements, while other quantities are more sophisticated like the *cosine similarity* or the *path distance* as introduced in Chapter 3. In any case, the matching quality has to be evaluated in terms of *accuracy* and *completeness* in order to retrieve a meaningful comparison between the different methods. Practically, so-called *retrieval* systems are used for this task. These frameworks are highly popular and have thoroughly been studied over the last decades.

Retrieval System   Given a large and unstructured collection of documents (e.g. text, images, video or audio files), a retrieval system returns those objects which are relevant to a certain search request initiated by a user. In contrast to the *data retrieval*, this search is based on semantic (instead of pure syntactic) characteristics. Therefore, the system provides a content-based analysis of these documents typically realised by performing a similarity search (cf. [Sch05]).

In content of this thesis, the system is exclusively operating on 2D and 3D objects. Thus, the object relevance towards a certain user request is determined based on the similarity of the pre-defined features between a *query* (input) and a *target* (observation) object.

> "Content Based Image Retrieval (CBIR) is a technique which uses visual contents, normally called as features, to search images from large scale image databases according to users' requests in the form of a query image." [SS10]

Since the retrieval system is purely used for the task of evaluation, only its core functionality has to be utilised. This encompasses the input interface, the features extraction, the similarity computation and the ranking mechanism of the output in descending order according to its similarity. Please recognise that a professional retrieval system accomplishes additional and more complex tasks. In order to obtain further insights to this topic, the reader is referred to [Sch05]. The remaining content of this section is devoted to the discussion of three indicators capable of rating the outcome of a retrieval system, namely the *recall*, the *precision* and the *average precision*. Given a data set $\mathcal{K}$ encompassing k objects, the entire test run over $\mathcal{K}$ would take k cycles. In each iteration, an object $\Omega_0 \in \mathcal{K}$ is selected as query and the system calculates the similarity between the query and all objects in $\mathcal{K}$. Once the test has completely been passed, the rating indicators are derived from its outcome. However, this generation process requires the existence of object labels informing the system about the objects' classes. Such a data set is called *Ground Truth*.

In general, the objects of the result list can be classified in terms of *true positive (FP)*, *true negative (TN)*, *false positive (FP)* and *false negative (FN)* in relation to the current query object. Let Ω_0 be the query coming from the object class $\mathcal{D}$ and assume that only the first m objects of the result list are taken into account. Then the object $\Omega_i \in \mathcal{K}$ with $i = 1, \ldots, k$ is classified as TP if $\Omega_{i,j \leq m} \in \mathcal{D}$, as TN if $\Omega_{i,j>m} \notin \mathcal{D}$, as FP if $\Omega_{i,j \leq m} \notin \mathcal{D}$ and as FN if $\Omega_{i,j>m} \in \mathcal{D}$ with j being the position inside the result list.

Recall   Measure for qualifying the amount of true positives retrieved by system in relation to the total amount of relevant objects in $\mathcal{K}$. This indicator is 1.0 if the result set consists of all relevant objects in the database.

$$R = \frac{\|TP\|}{\|TP\| + \|FN\|} \quad . \tag{2.16}$$

Precision   Measure for qualifying the amount of true positives in relation to the actual object list that has been retrieved by system. This indicator is 1.0 if the resulting output only encompasses objects which are relevant to the request of the user.

$$P = \frac{\|TP\|}{\|TP\| + \|FP\|} \quad . \tag{2.17}$$

Commonly, both values are generated for each cycle of the entire test run. The final indicators are then obtained by averaging all results. Keep in mind that the precision should not be interpreted without the recall and vice versa. As a simple example, assume that the system does always return the entire data set. Consequently, the recall would be 1.0 for every user request. Thus, the number has no relevance without its corresponding precision value. Moreover, a so-called *precision/recall graph* coupled with a *receiver operating characteristic curve* allows an easy interpretation towards the quality of the retrieval result. Nevertheless, for the purpose of comparison it is more convenient to specify the performance of the system by a *single* number.

(Mean) Average Precision   This measure is able to express the performance of the system by a single number. Instead of only considering the first m hits of the resulting object list, the entire output is taken into account by exploiting the order of the documents. In other words, the average precision summarises the shape of the precision/recall curve:

$$\int_0^1 p(r)\,\mathrm{d}r \quad , \tag{2.18}$$

where $p(\cdot)$ is interpreted as precision function over the recall r which is ranging from 0.0 to 1.0 (cf. [Zhu04]). In practice, this integral can be approximated by the following equation:

$$\sum_{k=1}^{N} P(k)\Delta R(k) \quad , \tag{2.19}$$

where N is the total number of objects in the data set, $P(\cdot)$ is the precision of the first k hits inside the (ranked) output list and $\Delta R(\cdot)$ the difference of the recall: $R(k) - R(k-1)$. Please notice that the average precision can only be calculated for a single request. In order to obtain a more sophisticated magnitude which qualifies the overall system performance, the *mean average precision* can be used. As before, this mean is generated by taking the average over all average precision results calculated during each query cycle. Another famous adaptation that is widely used in literature is the *interpolated average precision* as described in [Eve+10].

> "The intention of interpolating the precision/recall curve [...] is to reduce the impact of the "wiggles" in the precision/recall curve, caused by small variations in the ranking of examples." [Eve+10]

2.3   State-of-the-Art: Holistic View

This chapter discusses state-of-the-art methods concerning the subjects of *object representation* and *object matching*. Although both topics can be addressed independently, most methods try to approach them together. In consequence, this section does also neglect such a separation and covers both topics at the same time. Please keep in mind that there is a huge number of existing techniques. Hence, this chapter is not able to gather them all. Nevertheless, it provides a holistic overview of research activities over the last decades.

2.3.1 Shape and Object Recognition Approaches

Researchers realised early that the accuracy and the robustness regarding computer vision or machine learning tasks, e.g. object segmentation, detection or recognition can drastically be increased if 2D data is coupled with 3D (2.5D) information. Moreover, supported by the recent development of less expensive range-measurement devices (cf. Section 2.1), the number of research projects exploiting depth data has risen significantly. However, in the following a holistic overview of 3D *and* 2D state-of-the-art methods is given.

Boundary/Surface-Driven Approaches   The dominant set of works concerning the task of shape or object matching is operating on contours. In [SBC05] objects are categorised by local contour features coupled with a partially supervised learning technique. Nguyen et al. presents in [TNOL13] a local binary descriptor. Using this kind of description, the appearance of an object can compactly be encoded and used for human detection. The work of [Lu+10] introduces a contour model to represent a class of objects. This model is hierarchically decomposed into fragments. Afterwards, these fragments are grouped into part bundles and passed to a voting method. The object detection approach, published by Yang et al. in [YLL12], matches the contour parts of a model with the edge fragments previously localised in images. Therefore, correspondences are determined by finding dominant sets in weighted graphs. A partial shape matching approach for objects whose shapes are mildly affected by non-rigid deformations is published in [Cao+11]. The actual implementation is based on a Markov Chain Monte Carlo-based method. In [FTVG04], an appearance based object recognition system is proposed. The idea behind this method is to capture the relationships between multiple views of a model which are locally viewpoint invariant. Therefore, the authors introduce so-called region-tracks for the purpose of connecting image regions across multiple views. Moreover, Ferrari et al. [FJS10] propose a method for object class detection by training the underlying model based on a set of images. Instead of matching surface patches, the actual detection of objects is performed by localising shape boundaries employing a Hough-style-driven voting scheme. In [BMP02], the authors consider the shape of an object as an unorganised set of points. With the intention of setting up a reliable object recognition framework, they introduce a descriptor for establishing one-to-one point correspondences between two shapes based on their similarity. This descriptor became very popular over the last years and is better known as *shape context*. Its implementation exploits a histogram structure carrying all coordinate information of points relative to the currently selected reference. Similar to the shape context, [ML11] proposes another shape descriptor designed to perform a partial matching between a set of edge fragments and the contours of a target object. A further approach utilising the contour information for the task of object recognition is proposed by the authors Kass et al. [KWT88]. In their paper, they present a concept based on deformable splines. An elastic band, or snake, is iteratively evolved until it approximates the objects contour. Starting with a pre-defined shape, the spline is deformed more and more over the time. This evolution is influenced by internal and external forces or energies towards the object's contour. In [You98], this energy is monitored for contour matching. The method assumes that minimal energy is required to transform one contour curve into the other. In contrast to skeletons this representation type is not able to consider the shape's interior.

While the techniques above have been developed in the light of working in 2D, the following content introduces some methods operating in 3D. Therefore, it starts with a highly related approach presented by the authors Ma et al. in [MYL13]. Their method works on a set of 3D lines which have been extracted from a 3D point cloud. This extraction is realised by applying the Random Sample Consensus (RANSAC) approach iteratively to this point set. Once all lines have been generated, they are arranged inside a so-called affinity graph in order to identify Maximum Weight Cliques. The work of Nguyen et al. [NS09] introduces a similar approach for the extraction of 3D lines. Here point sets are determined by combining 2D images and 3D point clouds. Finally, these sets are divided into groups and an Eigen-analysis is performed. In [Sti+06], range data is used for object recognition. The authors propose a new Eigen-Curvature Scale Space method to extract reliable contour features which are then used to train support vector machines. Payet and Todorovic address the problem of view-invariant object detection and pose estimation [PT11]. By hiring view-dependent shape templates, the occurrence and the pose of an object is detected. The learning of such an contour-based object model relies on a huge amount of training examples. However, in the end, the process does only require a single image for the actual object detection. Drost et al. [Dro+10] introduce a method for recognising 3D free-form objects. Therefore, a global model description is generated based on oriented point pair features accommodated in 3D point clouds. The actual detection of 3D geometries is then realised by employing a fast voting scheme that matches these point pair features locally. The contribution of [RC+12] is also dedicated to 3D object recognition using 3D point clouds. The aim is to categorise objects using an adaptation of the well-known Speeded Up Robust Features (SURFs). This 3D modified version is computed locally on the object's surface taking into account the

3D point cloud data. Subsequently, a vocabulary of 3D visual words is generated. Spin images are introduced by Johnson and Hebert in their paper [JH99]. Their concept of surface description is highly popular in the context of 3D object recognition. Spin images are used to match mesh surfaces by finding individual surface point correspondences. Another method operating on local surface patches is presented by Chen and Bhanu [CB07]. Their descriptor is defined by a centroid, a surface type and a 2D histogram. The calculation of this descriptor is then limited to shape regions with large variations. The use of local 3D shape descriptors is also subject of [BN10] and [MBO06]. The researchers of [BN10] exploit the scale-variability towards geometrical properties to constrain the search space during the actual matching phase. In order to achieve this goal, scale-dependent corner-regions are detected as foundation for generating local and scale-invariant 3D shape descriptors. Mian et al. also analyse local 3D shape descriptors in [MBO06]. Aiming at this, they propose a multidimensional table representation consisting of multiple unordered range images, the so-called tensor. In [Tan+13], the local object geometry is expressed by histograms of oriented normal vectors (HONV). Therefore, 2D histograms are generated carrying the distribution of the *azimuthal* and the *zenith* angle over a certain set of normal vectors. Global as well as local free-form surface features have been investigated comprehensively in the past, an overview up to the year 2001 is given in [CF01].

Skeleton-Driven Approaches   Skeletons are popular contour-based representations and are of special interest in this thesis. A definition as well as their role in the area of computer vision (and graphics) is subject of Section 2.4.1. Apart from this, the present paragraph is concentrating on the state-of-the-art methods using this kind of representation. The skeletons or rather the medial axes have been first proposed by Harry Blum in the late 1960 [Blu67b]. In subsequent years, skeletons have deeply been investigated and further methods have been proposed to extract their structure or to exploit their properties for object recognition.

In [OI92; OK95], skeletons are generated based on Voronoi diagrams. Here the resulting skeleton is formed by a subset of Voronoi edges coupled with a hierarchical clustering of the skeleton branches. Baseski et al. introduce in their paper [BET09] a tree-edit driven shape matching. Therefore, the skeleton characteristics reflecting the properties of the shape are mapped to a tree. The dissimilarity between two shapes is then computed based on the tree-edit distance. Highly related to this approach is the concept of so-called shock graphs or shock trees [SK96]. Shocks are singularities of the grassfire evolution and occur whenever two or more fronts collide. A shock graph, in turn, is a description of a skeleton in form of a directed acyclic graph. Shock graphs are also utilised in [Kle+00; KSK01], where a method is proposed aiming at shape comparison. Finally, the similarity of two shapes is calculated by summing up the cost of deformations needed to transform one shock graph to the opponent one. The deformation costs between two shock transitions is also called edit distance and each edit operation is associated with a pre-defined cost. Shock graphs are also exploited in context of a view-based 3D object approach as presented in [Mac+02]. In [BL08], a matching algorithm with the name "Path Similarity Skeleton Graph Matching" is proposed. This method starts with computing the matching costs for each pair of skeleton end points using the concept of shortest paths. In a second step, the costs are passed to the Hungarian method and the final matching as well as the overall similarity is retrieved. This technique is also subject of a deeper discussion in [Hed+13]. Xu et al. also employs shortest paths in [XWB09]. Motivated by the work of [BL08], they are utilising all shortest paths between *all* pairs (end and junction points). This extension is encouraged by the fact that junction points carry global properties of the shape's structure. In [Bai+09], a further skeleton-based object detection approach is introduced. While the skeletal structure is capturing the main characteristics of the object, its branches are respectively assigned to a set of boundary-part templates. Then, a tree-union structure is learnt on this data in order to perform a rapid object detection. Further approaches which are using both the contour and the skeleton of a shape, are presented in [BLT09] and [Zen+08]. Strictly speaking, [BLT09] combines the properties of the contour and the skeleton for the task of shape classification, whereas the authors in [Zen+08] employ the same information to implement a meaningful shape decomposition. From a global view on the topic of object recognition, it is highly essential to represent the object's shape adequately. In addition to this, the selection of the right matching approach is equally important to build up a well-operating system. While most of the methods enforce a one-to-one matching, the authors of [Dem+06] and [DSD09] propose a many-to-many alignment using medial axis graphs. Therefore, the nodes of two graphs are first embedded into the same fixed-dimension Euclidean space in order to achieve a higher level of shape abstraction. Subsequently, the many-to-many matching is enabled by employing the Earth Mover's Distance (EMD) [Dem+06]. An extension of this approach is presented in [DSD09] where the focus is set to the medial axis graphs. The work in [Yan+09] aims at the generation of stable skeletons by utilising the Discrete Curve Evolution (DCE) (cf. Section 3.1.1) to determine a small set of salient contour points. The actual skeleton branches are then established by determining maximal disks inside the shape.

In [PK98] and [PK99] two thinning approaches are presented to generate curve skeletons of elongated 3D binary objects. As usual for thinning algorithms, the medial axis is extracted by shrinking the object's surface iteratively. Here this process is implemented by executing sub-iterations in parallel. A further 3D thinning approach is discussed in [MS96]. In analogy to the previous one, this technique also works in parallel while it preserves the connectivity of the resulting structure. 3D Skeletons can also be obtained by Voronoi diagrams like in [HBK01]. The skeletal shapes are extracted by a three-dimensional Voronoi technique applied to polygonal surfaces. Another skeletonisation approach is proposed in [Ren09]. It takes a special role in context of this thesis and thus, it is subject of a deeper discussion in Chapter 7. Using the information from a distance map, an importance measure is established to determine so-called Jordan curves. These curves are then exploited to extract a curve skeleton for the purpose of segmentation. The generation of curve skeletons is also addressed in [Sha+07]. Therefore, the authors introduce a deformable model that evolves inside the object to approximate the corresponding shape. Afterwards, the skeleton points are localised by the centre curve of the evolved model. Other deformable model-driven approaches are realised based on level sets (cf. Section 6.1.1), e.g. in [VUB07]. The authors of [HF05; HF07b] present a framework to compute the object's centre line by tracking the evolution of wave fronts over time. The paper of [Au+08] is also focusing on the extraction of curve skeletons using a mesh contraction process. The object's mesh is iteratively contracted by a global positional constraint Laplacian smoothing. An extension of this is published in [Cao+10] capable of taking point clouds instead of meshes. Moreover, the method is still applicable in presence of missing data, e.g. incomplete range scans. Cornea et al. [Cor+05b] introduce the concept of hierarchical curve skeletons. By exploiting a repulsive force field, topological characteristics of the skeletal structure are identified.

In addition to this, the paper [SK05] of Sebastian and Kimia provides a comparison of boundary-based shape representations including shock graphs. Their work clearly shows that skeletons are computationally more complex than other shape-based representation types. Another skeleton graph-driven approach is proposed in [MBH12]. Similar to the shock graph, a so-called Reeb graph is used to encode topological as well as geometrical properties of the 3D shape. With access to this data, the authors propose a novel graph matching technique that operates on the shortest paths between the skeleton end points. In [BI04], a method for matching 3D polygonal geometry models is motivated. Aiming at this, skeleton graphs are generated which carry the overall shape information. The goal is then to find the largest common sub-graph in order to estimate the similarity of two 3D models. In [Cor+05a], the authors encourage the use of skeletal structures to perform the object retrieval based on a many-to-many matching approach that employs the EMD to compute the dissimilarity of curve skeletons. Although the execution of the EMD requires high computational efforts, the method is able to achieve promising results. In [Sun+03], two skeleton graphs are matched by exploiting node signatures characterising the structure of the node's underlying sub-graph. Subsequently, the similarity of two nodes is defined by the distance between their signatures. Another skeletal representation type are so-called surface skeletons. Please keep in mind that surface skeletons are not in the scope of this thesis. However, for the sake of completeness, a brief overview of related works is given in the following by starting with the one of [Zha+05]. Here medial surfaces are interpreted as DAG for the task of indexing and matching 3D objects whose shapes are affected by articulation. However, it has been shown that their approach for matching medial surface graphs is able to achieve excellent results, particularly for objects with articulating parts. In [Hay+11], surface skeletons are utilised for the similarity computation. Given a voxelised 3D input structure, a histogram is generated as an indicator for the feature distribution. Finally, this descriptor is passed to the similarity calculation process of two 3D shapes. The analysis of surface skeletons is a non-trivial task and the interested reader is referred to [SP08] in order to find a good entry point for this topic.

Miscellaneous Approaches   In [LF09] a method is proposed using Google's 3D Warehouse objects for the purpose of training a classifier. The benefit of this database is its huge amount of labelled training instances. Subsequently, this classifier is evaluated in a robot navigation setup capturing 3D real-world objects with a laser scanner. Connected to this, the work of Lai et al. [Lai+11a] presents a large RGB-D data set consisting of 51 categories encompassing 300 objects. For each object instance multiple views from varying perspectives are captured. Additionally, a comprehensive evaluation is performed by state-of-the-art methods. In [Lai+11b], the authors propose a sparse distance learning approach for object category and instance learning. Therefore, they present a novel view-to-object distance learning technique. The actual implementation is based on a weighted combination of feature differences between multiple views. Another 3D object detection concept is introduced in [SA12] using an adaptive fusion scheme for RGB-D data. Strictly speaking, a two-tier architecture of local experts with two different gating functions weights the confidence of each detector and adjusts the relative importance of each sensory cue with respect to the other one. Another adaptive object recognition mechanism is proposed in [LS13] based on spatio-temporal features which are automatically extracted from RGB-D data. The actual feature learning process is then considered as an optimisation problem. A comprehensive evaluation attests the method an excellent recognition performance. In contrast to the previous approaches, a technique for multi-view object class detection is published in [LS10]. The working principle behind this method considers the learning process of the object's appearance and its geometry as two separated tasks using different databases of training data. While the appearance properties are learnt on real-world images, a synthetic (CAD) model encodes the 3D geometry of an object. Finally, both types of information are linked together enabling 3D pose estimation. Yan et al. propose an object class detection method using multiple 2D training instances fused with properties of the 3D objects [YKS07]. Spatial connections are then established between multiple 2D views by mapping them to the surfaces of the 3D models. Afterwards, the method localises correspondences of a given 2D target image to the previously generated 3D feature models. Apart from this, Bo et al. [BRF] introduce a kernel-driven approach for object recognition using depth data. The corresponding kernel features encompass the size, the shape and the depth edges of an object. The actual performance of these features is then demonstrated by performing a comprehensive object instance and category recognition analysis. In [BRF12], the same authors present an unsupervised learning approach using a *hierarchical matching pursuit (HMP)*. In more detail, dictionaries are first generated for different data channels based on pixel patches (first layer). Subsequently, a second layer compresses the previous one to sparse linear combinations of code words. Similar to this method, Li and Guskov propose in their work [LG07] a recognition concept implementing a pyramid kernel function to calculate the similarity between two depth images. Therefore, the function takes a set of local shape descriptors derived from the surface patches of the object's shape.

2.3.2 Vessel-based (Matching) Approaches

After a more generic view on object recognition, this section now switches to a more concrete application domain, namely the processing of vessel structures. This separated consideration is reasonable in order to support a better access to the project of Chapter 6. Moreover, the delimitation of the subject becomes more obvious and generates its own thematic unit.

A plethora of works in this area is devoted to the segmentation process of the vessel structure. An automatic segmentation approach is presented in [dBr+02] based on an active shape model (ASM). The actual fitting of the ASM is performed on sequential slices. In each iteration, the previously obtained contour is used as initialisation for the next cycle. In [Zhu+06], a level set-driven segmentation method is exploited to refine a roughly sketched initial surface. Therefore, two external analysers are taken into account operating on the global region and the local features. In analogy to this, another level set technique is proposed in [Zha+08]. Since the method is of special interest for this work, it is subject of a deeper discussion in Chapter 6. A review of 3D vessel lumen segmentation concepts is given in [Les+09]. The review discusses state-of-the-art approaches in terms of modelling and feature analysis. Another review towards the topic of vessel extraction is given in [KQ04].

With access to the segmentation result, the vessel can be processed for the purpose of medical diagnosis. In [MLP11], Shun et al. propose a 2D/3D registration method taking into account the pre-operative 3D aorta model as well as the intra-operative 2D x-ray images. The main contribution of this work is its novel registration principle requiring only one instead of two images. The idea of combining 2D with 3D is also investigated in [Rah+10]. Here Raheem et al. employ a rigid 2D/3D registration system for the task of merging CT data with fluoroscopy images. In more detail, they deform the CT-based model surface with the intent to display the interventional scene. In [Lia+10], a graph is generated from the 3D CT volume and, in addition to this, a 2D distance map based on the x-ray images. Finally, the graph similarity is calculated using the underlying structure of the distance map. A further 2D/3D registration technique is motivated in [Du+12]. Here the authors tackle the problem of matching fluoroscopy data with pre-operative volumetric structures. Their contribution primarily aims at the automatic movement compensation during that registration process by detecting the pelvis boundary.

The almost tubular shape of a vessel is highly suitable for the task of representing its structure by a skeleton. Thus, a 2D skeletonisation method is described in [YCS00] for determining vessel paths and branching patterns, whereas [VUB07] introduces a 3D approach operating on volumetric Computed Tomography Angiography (CTA) series (cf. Chapter 6).

2.4 State-of-the-Art: Detailed View

After inspecting the wide field of object representation, matching and recognition, this content part takes a deeper look at the working principles towards a set of selected state-of-the-art methods. These techniques are closely related to this work since they are employed in two or more of the following projects. Beginning with skeletons as a candidate for object representation, matching methods conclude this section.

2.4.1 Skeletons: Definition, Extraction and Pruning

> "[Skeletons are the transformation of] a shape into another one that is of a lower dimensionality than the shape it describes [...]. The skeleton of a 2D shape can be seen as its stick-figure representation" (cf. [Ren09]).

The concept of skeletons has been introduced in 1967 by Harry Blum who proposed the so-called *Medial Axis Transform (MAT)* for the purpose of biological image analysis [Blu67a]. As the name already suggests, the idea is to reduce an object Ω to its *medial axis* as shown in Figure 2.11.

Medial axes or skeletons represent geometrical characteristics of a shape in terms of topology and geometry (in a compressed form). The skeletal structure in 2D, and most often in 3D, is a composition of curves, the so-called skeleton branches. Each branch typically has one end point that connects to a junction point. Junction points, in turn, are places inside

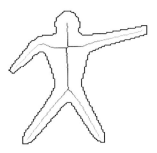

Figure 2.11: Exemplary instance of a medial axis.

a skeleton where three or more branches meet. A widely accepted definition of the medial axis is the so-called *grassfire propagation model*:

> "Imagine a fire lighted on the shape boundary and the fire front moving at unit speed inward, in normal direction of the front. Then, the skeleton is defined as those points where the fire front meets itself." [Ren09]

This continuous movement of the fire front can be formulated based on the following partial differential equation:

$$\frac{\partial C(\mathbf{x}, t)}{\partial t} = -\alpha \mathbf{n}(\mathbf{x}) \quad , \tag{2.20}$$

where C depicts the wave at time t, $\mathbf{n}$ is the normal vector on unit scale oriented outwards to the moving fire front with α as constant speed of the movement. Both sides of the equation are parametrised with $\mathbf{x}$ (cf. [SP08]).

Another definition operates on bi-tangent disks. Therefore, each interior point of the object is considered as the centre of a disk. Consequently, the medial axis only encompasses those (bi-tangent) disks with maximum size or, in other words, which are entirely contained inside the object's shape. These medial axis points are called *loci* and are defined as tuple $(\mathbf{p}, r)$, with $\mathbf{p}$ being the centre of the disk with radius r (cf. [SP08]). Please notice that the information of r determines the intensity of the grey values in Figure 2.11. In addition to this, it is worth knowing that r can also be viewed as the time which is required by the grassfire front to reach the point $\mathbf{p}$. Figure 2.12 provides a further example of a medial axis but this time, the figure emphasises the inscribed disks determining the set of loci.

Even though both terms, the medial axis and the skeleton, are frequently used to describe the same structure, the MAT does not produce skeletons according to its proper definition. The medial axis, e.g. carries the radial distance in contrast to skeletons. However, this fact is often neglected in literature and likewise it is done in this work. Henceforth, the terms are used interchangeably. For the continuous case, the *skeleton* (expressed by the symbol S) can be formulated mathematically as follows:

$$S = \{p \in \Omega \mid \exists a, b \in \partial\Omega, a \neq b, \|p - a\| = \|p - b\| = D(p)\} \quad , \tag{2.21}$$

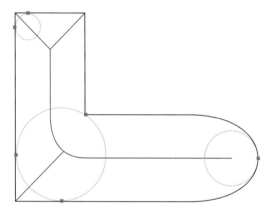

Figure 2.12: Exemplary instance of a medial axis (red) together with three maximum inscribed disks (drawn in green).[12]

with D being the distance transform (DT) of an object as introduced in Section 2.2.6 and the points a, b on the object's boundary $\partial\Omega$. This definition is revisited in Chapter 7, where a, b are better known as *feature points*. Basically, Equation (2.21) just says that a skeleton is formed by a set of interior shape points satisfying the property of having at least two boundary locations at minimum distance (cf. [Ren09]). While two features points constitute the lower bound for being a member of the skeletal structure, the actual amount of points (touched by the disk) can be exploited for a more sophisticated classification. In total, three classes of skeleton points are distinguishable, namely *junction points*, *end points*

---

[12]http://commons.wikimedia.org/wiki/File:Medial_axis_example_2d.svg, [online: 19th August 2015]

and ordinary *skeleton points* forming the connection or the skeleton branch between the two former ones. The ordinary skeleton point hits exactly two boundary fragments at minimum distance, whereas a junction point yields at least three of them. In contrast to this, the end point of a skeleton branch typically constitutes a continuous set of feature points. This unbalanced relation arises by the fact that an end point is usually located at or close to boundary sections of convex curvature maxima as illustrated in Figure 2.12. Having this in mind, it is no surprise that the maximum disk partially coincides the object's shape [Ren09]. This classification scheme is also applicable in 3D. However, it should already be stressed that the actual detection of these points, known as *skeletonisation*, embodies a non-trivial task in 2D and 3D.

Definition 2.4.1. *Skeletonisation is the process of reducing an object to its skeleton.*

Definition 2.4.2. *A skeleton is an abstracted representation of an n-dimensional object ($\Omega \in \mathbb{R}^n$) reduced to a less complex structure. This structure is a connected set of $(n-m)$-dimensional elements embedded in the same space as Ω.*

Commonly, the degree of abstraction (m) equals one, e.g. in 2D space. In 3D, this situation varies a bit due to the fact that two abstraction levels are available. On the one hand, the 3D object can be reduced to a set of 2D surfaces ($m = 1$) forming the *surface skeleton*. On the other hand, a 1D curve representation ($m = 2$) does also exist, the *curve skeleton*.

Skeletons in 3D - Surface versus Curve Skeletons Both representation types are widely studied with the intent to represent 3D objects by emphasising their most characteristic properties. While curve skeletons exhibit a higher visual similarity to skeletons in 2D, surface skeletons, or medial surfaces, are higher related to the concept of the MAT. Thus, the definition of the MAT can easily be mapped to 3D by substituting the 2D disks by 3D balls.

"Let $[\Omega]$ be a connected closed set in $\mathbb{R}^n$. A closed ball $B \subset \mathbb{R}^n$ is called a maximum inscribed ball in $[\Omega]$ if $B \subset [\Omega]$ and there does not exist another ball $B' \neq B$ such that $B \subset B' \subset [\Omega]$." [SP08]

The term *surface* is a bit misleading since the resulting skeletons might be a composition of 2D manifolds *and* curves. It strongly depends on the upper-level object geometry which primitive is generated as representative for a certain region. While surfaces are the result of areas with a flattened form, curves are the outcome of cylindrical shape parts. Nevertheless, the structure of a curve skeleton is more intuitive and easier to handle. Strictly speaking, curve skeletons supply attractive computational properties towards, e.g. the detection of end points or the calculation of branch lengths. Figure 2.13 demonstrates both types on the example of a 3D hand object. In a consequence of this, curve skeletons are more widespread than the surface skeletons. Nevertheless, the extraction of curve skeletons is equally sophisticated and actually suffers from the absence of a generally accepted definition. According to [SP08], a curve skeleton $S^{3,1}$ is a subset of an object Ω if (i) $S^{3,1}$ has the same topology as the object, (ii) the skeletal structure is centred inside Ω and (iii) $S^{3,1}$ is one-element thick. Please notice that the skeleton notation is realised in analogy to [Ren09], where the superscripts of the skeleton symbol are carrying the information about the object's dimension n and the degree of abstraction m. In addition to this, the curve skeleton is intensively discussed in Chapter 6 and Chapter 7.

The classification of 3D skeleton points is similar to the process introduced for 2D. The surface skeleton $S^{3,2}$ as well as the curve skeleton $S^{3,1}$ consist of junction, end and skeleton points. While the given classification scheme can be mapped immediately to the structure of a curve skeleton, it is slightly adjusted for $S^{3,2}$. In other words, an ordinary skeleton point is located on the 2D manifolds, whereas a junction point typically occurs on a line (together with other junction points) indicating the intersection of multiple surfaces. The skeleton end points, in turn, are located on the rims of the single surfaces.

Figure 2.13: Illustration of two valid skeleton representation types generated based on a 3D hand geometry (cf. [Ren09]). (**Left:**) A surface skeleton composed by surface structures (palm) and curves (fingers). (**Right:**) The corresponding 1D curve skeleton.

In the following, a brief overview of the most desirable skeleton properties is given for both 2D and 3D (cf. [CSM07; Ren09]). Afterwards, concrete skeletonisation methods as well as the concept of skeleton pruning ares discussed.

Centred   This property is already given by the skeleton definition specifying that its structure has to be locally centred inside the object. As trivial as this property might sound, the corresponding implementation turns out to be challenging. Discretisation artefacts in general as well as the degree of freedom towards the centre detection in 3D exacerbate this task [Ren09]. One way to align the 1D skeletal structure uniquely in 3D is realised on the medial surface [CSM07].

Homotopic   The skeleton reflects exactly the topology of the object. This implies that the skeleton has the same number of connected components, tunnels and cavities as the original object (cf. [CSM07]). This property only holds if the object's structure does also fulfil the condition of being connected.

Connected   All skeleton points are required to be adjacent (at least to one of the others) in order to form a single structure without any gaps. The property is implicitly given if the skeleton is homotopic.

Transformation invariant   The skeleton structure is invariant to isometric transformations since they do not change the shape's geometry, e.g. translation or rotation, and thus, the distance between two points remains equal. This property is not always valid inside the discretised space.

Thin   This property basically means that the skeleton representation has to collapse in one dimension to an $(n - m)$-structure. In other words, it forces the skeletal structure to have a thickness of one element. However, if the curve skeletons this formulation has to be adapted slightly.

Robust   Like all kinds of representation, it is of particular importance that their extraction works reliable for the purpose of analysis, e.g. the matching. In the scope of skeletons, this reliability encompasses *uniqueness*, insensitivity against *noise* and, associated with this, the absence of *spurious skeleton branches*. Please bear in mind that these characteristics do only sketch a rough outline.

It is worth knowing that these conditions are formulated in the light of continuous shapes. However, their maintenance in a discretised environment cannot be guaranteed due to *discretisation artefacts* (cf. Chapter 7). In summary, skeletons are powerful but difficult to handle structures. Hence, the following content is dedicated to the methods capable of generating skeletal structures.

2.4.1.1 Skeletonisation Approaches

In the following, the focus is set to the most famous skeletonisation techniques in this area, namely (i) methods based on *distance maps* (ii) *thinning* and (iii) *Voronoi diagrams*. Although there is a plethora of other approaches to produce skeletons, like the *iterative least squares optimisation* [WL08], most of them are developed on top of these core concepts.

Distance Transform   This operation can be employed to generate a distance map or field. Distance maps play an important role in the field of skeletonisation algorithms since the skeletal structure is derivable from them. Frequently, this transformation is combined with other methods, e.g. in [Don+03], where the classical thinning approach is steered by a distance map. The distance transform is also subject in Section 2.2.6, Chapter 6 and Chapter 7.

Basically, a distance map has to be understood as a scalar-valued field. Strictly speaking, given an arbitrary binary input structure $\mathcal{K}$ with a labelled subset of lattice points $\mathcal{D} \subset \mathcal{K}$, the resulting distance map is carrying the distance of each point $\mathbf{p}_i \in \mathcal{K} \setminus \mathcal{D}$ to its closest element in $\mathcal{D}$. A concrete example might be the situation, where $\mathcal{K}$ represents a 2D binary image displaying the boundary of a shape. In this case, the subset $\mathcal{D}$ is formed by the points of the boundary and the distance map encodes the distance of each pixel to its closest boundary section. These values have to be generated iteratively starting at the given set $\mathcal{D}$ labelled with zero distance. While a good introduction can be found in [Fab+08a] discussing multiple distance transform algorithms, Figure 2.14 shows an example of a distance map and its corresponding skeleton.

Figure 2.14: Example of a distance map-driven skeletonisation approach. (**Left:**) Contour of a person, (**Centre:**) Distance map generated based on the contour information illustrated on the left. Please notice that the distance information is given by intensity values, where a dark colour corresponds to a high distance. (**Right:**) The skeleton derived from the distance map by tracing its peaks and ridges.

For the purpose of skeletonisation, the distance map can be considered as a *height map* with peaks and ridges. The desired skeleton structure is then obtained by tracing these local maxima based on an uphill climbing approach. Mathematically speaking, a *steepest ascent method* has to be employed to connect these singularities (non-differentiable locations inside the distance map) in order to obtain a connected and centred skeleton (cf. [Ren09] and [SP08]). Distance Maps also constitute an adequate tool for determining the loci of maximum disks and balls. Finally, it shall be stressed that the underlying complexity and accuracy are correlating to each other. By replacing the Euclidean distance function, e.g. with the Manhattan metric, the complexity can be decreased at the expense of a lower accuracy.

Thinning   This class of algorithms is widely accepted as an appropriate solution to the task of reducing an object to its skeleton. Starting at the outer rim of the object's shape or body, the process iteratively removes grid points with the intent to shrink it. This process repeats until a thin skeletal curve or surface is obtained. The concept behind the thinning stays the same no matter if it is applied in 2D or 3D. Thinning algorithms are not able to operate on every 3D representation type, only voxelised data structures are compatible to them. From a more general perspective, thinning is a morphological operation using so-called structuring elements which guide the entire process. These elements can be viewed as a template controlling the deletion (or addition) of grid points. Figure 2.15 illustrates the principle of thinning. For more information, please refer to [GW06].

Figure 2.15: The figure illustrates the working principle of thinning approaches in general. **(Left:)** The input image. **(Centre:)** One further exemplary iteration step. **(Right)** The final skeleton.

In context of the skeletonisation topic, the actual design of the structuring element has to be carried out carefully to preserve the object's topology. In 3D, a bunch of elements has to be employed in consequence of the higher dimensionality. However, due to discretisation artefacts, the extraction of the skeletal structure often conflicts with the properties stated above. Moreover, induced by the local character of thinning approaches, the geometrical information about the overall shape is entirely discarded. Thus, the detection of end points runs the risk of having a high false positive or false negative rate. In addition to this, the end points have to be kept during the thinning to prevent a shape collapse leading to a single point [Ren09]. Further potential and negative side effects of this locally driven operation are unconnected skeleton branches and a non-deterministic peeling behaviour.

Strictly speaking, changing the order in which lattice points are removed from the boundary might lead to varying results. In literature, the problem is faced by parallelising the process of the erosion as described in [WL08]. Moreover, thinning algorithms tend to produce spurious skeleton branches requiring the application of a pruning method as a mandatory task in many cases. Summarised, it shall be figured out that thinning approaches are powerful and well performing algorithms towards the task of skeleton extraction. Nevertheless, their attractive computational properties in terms of speed and memory usage have to be dismissed for the purpose of retrieving robust and accurate structures. Most often a higher accuracy can only be realised by incorporating the global shape information. For more detailed information, please consult [LLS92] and [SP08].

Voronoi Diagrams   The decomposition of a space into regions, the so-called *Voronoi cells*, has widely been used in various applications, e.g. in [Gar+06], properties of the Voronoi diagram are used for mobile robot navigation. Thus, it has been analysed and studied intensively, especially in the context of *computational geometry*. The actual diagram is obtained as follows: Given a subset of points in an n-dimensional space (with $n = 2, 3$), also known as *sites*, Voronoi cells are generated for each site by gathering all remaining points having the closest distance to this site than to all others. These cells can be viewed as convex, not always finite, polygons (cf. [SP08]). An example of a Voronoi diagram is shown in Figure 2.16, where the sites are marked as red squares.   As part of the skeletonisation processes, it is worth knowing that the Voronoi edges (the separating lines between two adjacent cells) are of the same distance to both sites which are responsible for their existence. Having this in mind, it is no surprise that the skeletal structure can be obtained by passing some of the boundary points, declared as sites, to a Voronoi diagram generating algorithm. Thus, the boundary of an object has to be sampled sophisticatedly in order to retrieve a representative subset of the present shape. Like every other sampling procedure, its implementation is a challenging task. The more points are selected for the purpose of skeleton extraction, the more branches are generated. Even though the accuracy of the skeleton can be improved this way, the computational demands are unattractive. In addition to this, a higher amount of Voronoi edges also requires a more complex strategy to remove spurious skeleton branches (cf. Section 2.4.1.2). Connected to this, the diagram generation reacts sensitive to boundary

---

[13]http://www.dma.fi.upm.es/mabellanas/tfcs/fvd/voronoi.html, [online: 19th August 2015]

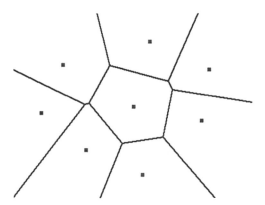

Figure 2.16: Example of a Voronoi diagram[13].

noise resulting in further falsely determined Voronoi edges. However, once these problems have been tackled, the actual skeleton is formed by a subset of all existing Voronoi edges except for those which are residing outside the object's boundary.

Please further recognise that the Voronoi diagram is similar to the grassfire propagation model but varies in the fact that only the sites are set on fire instead of the entire boundary section. Hence, the fire is not moving as a wave but as circular fire fronts until they meet on the Voronoi edges (cf. [SP08]). Voronoi diagrams are difficult to generate on discretised data structures, e.g. pixel or voxel grids. One way to alleviate potential problems is to involve a sub-grid point accuracy. Furthermore, the Voronoi diagram corresponds to the dual graph of the *Delaunay triangulation* and can be derived from it[14].

Please bear in mind that other skeletonisation methods can be found in today's literature. Some of them are based on novel concepts but most are operating in analogy to the principles stated above or extend them (slightly). A compact overview is given in Section 2.4.1. However, all of them are sensitive to boundary noise affecting the structural accuracy of a skeleton. This concerns both 2D and 3D approaches to the same extend with the result of anomalies in form of *spurious branches*. These falsely positive classified curves or surfaces represent redundant or not relevant shape information. The procedure of eliminating these outliers is called *pruning* and is going to be discussed in the next section.

---

[14]http://mathworld.wolfram.com/VoronoiDiagram.html, [online: 19th August 2015]

2.4.1.2   Skeleton Pruning

As mentioned above, skeletonisation algorithms are suffering from an inherent sensitivity to boundary noise. This kind of shape perturbation can already arise during the image acquisition, the object extraction or the shape's boundary sampling. Additionally, certain affine transformations applied to discretised structures might also be responsible for noise. In consequence of this, skeleton branches are generated as a result of this noise or, in other words, which are not relevant for the overall shape impression. This behaviour, in turn, conflicts with the expectations on a skeleton of being a *robust* and *accurate* representation of the upper level shape.

Having this knowledge in mind, the general idea of pruning is reasonable and aims at the removal of these spurious skeleton branches. Pruning constitutes one of multiple possibilities to remove falsely detected elements in a skeleton. While the pruning is attached as a post-processing step to the actual skeletonisation process, other methods take part in the pre-processing or operate in parallel to the extraction of the skeleton. An example for each instance is given in [POB87] and [VUB07], respectively. Nevertheless, the use of pruning algorithms with the objective to delete spurious skeleton curves or surfaces dominates literature. Please notice that the notation in this thesis is realised in analogy to [Ren09] which means that the original skeleton is depicted by the symbol S, whereas the pruned structure is marked with an additional pruning parameter τ: S_τ.

Although pruning approaches are an appropriate instrument for the task of skeleton cleaning, they are limited to deletion operations. This implies that the topological structure of the pruned skeleton equals the outcome of the same object *without* noise. Hence, S and S_τ are almost identical towards their number of branches, end and junction points, but not necessarily identical according to their alignment of curves or surfaces. As illustrated in [Ren09], Figure 2.17 shows such a situation. Even if the pruning removes the two additional branches successfully, it is not able to restore the structure of the skeleton completely. However, for the majority of applications, this result would be acceptable. The actual decision whether a branch shall be removed or not is controlled by the so-called *importance measure*. This measure assigns a value to each skeleton point indicating its relevance to the structure of the object's geometry. The higher the influence of a skeleton point towards the shape appearance, the higher the importance value which is assigned to it. Nevertheless, finding a meaningful and appropriately operating importance measure is a difficult task. Concrete examples typically exploit geometrical properties of the object, e.g. the angle between two feature vectors [Blu73] or the length of the shortest boundary segment between two features of a certain skeleton point [OK95]. Other approaches monitor the area/volume of the protrusion in which the skeleton branch ends [TH02] or calculate the *bending poten-*

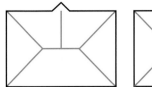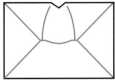

Figure 2.17: The figure demonstrates the limitations of the pruning process. **(Left:)** The original shape without boundary noise and its corresponding skeleton. **(Right:)** The same shape as before but this time it is slightly changed on its top. Even if the pruning removes the two additional branches successfully, it is not able to realign or to add skeleton curves.

tial ratio [She+11]. Further information about pruning, importance measures and pruning strategies can be found in [SB98; Ren09; SP08] as well as in Chapter 3 and Chapter 7.

2.4.2 Graph Matching Techniques

Graph matching algorithms are widely used for plenty of applications like the assignment (cf. Section 2.2.1) and the shortest path problem (cf. Section 2.2.5). Keep in mind that there are more real-world as well as research scenarios which intensively exploit the graph representation (cf. Section 2.2.2) and, of course, graph matching techniques. However, a comprehensive discussion is prohibitive for the scope of this thesis.

Thus, the following content is exclusively dedicated to the most related graph matching approaches in context of this thesis. The talk is about the well-known Hungarian method and a rather new method determining Maximum Weight Cliques (MWC) inside a given graph. While the former is limited to complete (one-to-one) matchings, the latter allows the establishment of partial correspondence configurations.

2.4.2.1 The Hungarian Method

The Hungarian method, also known as Kuhn-Munkres algorithm, was first developed in 1955 by Harold W. Kuhn [Kuh55]. Later, in 1957, it has been revised by the researcher James Munkres who pointed out that the algorithm is capable of solving the classical $n \times n$ assignment problem within polynomial time.

As already stated in Section 2.2.1, the classical assignment problem can be expressed by a complete bipartite graph, where the vertices represent the entities to be matched. This weights at each edge indicate the costs for matching two vertices. The goal is to find a (perfect) complete matching with minimum weight. The matching costs can easily be calculated by summing up all edge weights being part of the final correspondence configuration $\mathcal{M}$:

$$\psi(\mathcal{M}) = \sum_{e \in \mathcal{M}} \psi(e) \quad , \tag{2.22}$$

where $\psi(\cdot)$ returns a cost value in relation to its input which can be the matching ($\psi(\mathcal{M})$) or a single edge ($\psi(e)$). In order to find the matching with minimum weight: $\operatorname{argmin} \psi(\mathcal{M})$, the Hungarian method is used.

In the following, a rough introduction to its working principle is given. To obtain further information, please refer to [Kuh55]. For the time being, it should be known that the edge weights (or costs) have to be arranged in an $n \times n$ (cost) matrix **C**. Based on this matrix, the algorithm starts to generate so-called *zero-elements* in the columns and rows of **C**. In a next step, the *minimum* number of rows and columns needed to cover all zero-elements has to be determined. If this number equals n (the maximum column/row spread) a unique matching is found. Otherwise, the algorithm creates iteratively new zero-elements until the condition above is fulfilled. Bear in mind that this concept does only work under the assumption that a valid matching configuration exists.

Moreover, the Hungarian method enforces a one-to-one matching. However, this property is undesirable for a steadily growing number of applications requiring the capability of finding partial matchings, e.g. between two shapes with a varying number of feature points. Unfortunately, there is no native implementation of Hungarian method able to satisfy this demand. Instead, it is common practice to exploit so-called *dummy nodes* as latent (virtual) pairing partners with the intent to enable a complete one-to-one matching. The amount of dummy nodes is equivalent to the difference between both dimensions of the cost matrix in order to square its shape. During the actual matching, the dummy nodes are then assigned to those entities which do not find any proper correspondence. Nevertheless, the estimation of appropriate cost values, which have to be attached to each dummy node, has to be considered as a non-trivial task (cf. Chapter 3).

2.4.2.2 Maximum Weight Cliques

In [ML12], the authors propose a more general approach to find an appropriate solution to a broad range of applications facing the assignment problem. Therefore, the correspondence problem is formulated as an *integer quadratic program* operating on *quadratic equality constraints* (cf. Section 2.2.4). Furthermore, this working principle can be mapped to any combinatorial optimisation problem which is modelled as stated above, e.g. segmentation jobs. The most challenging tasks in this context are: (i) The appropriate population of the affinity matrix **A** or the correspondence graph $G^\star$ with weights assigned to all vertices/edges and (ii) the detection of a meaningful set of domain constraints. Once both quantities are available, the problem is formulates as follows:

$$\underset{\mathbf{x}}{\arg\max} \quad g(\mathbf{x}) = \mathbf{x}^{\mathsf{T}} \mathbf{A} \mathbf{x} \quad \text{s.t.} \quad \mathbf{x}^{\mathsf{T}} \mathbf{M} \mathbf{x} = 0,\ \mathbf{x} \in \{0,1\}^n \quad , \tag{2.23}$$

where $\mathbf{x}$ depicts the n-dimensional selection vector towards all vertices in $G^\star$ and $\mathbf{A}$ its corresponding $n \times n$ affinity matrix with $\forall i, j = 1, \ldots, n : \mathbf{A}_{i,j} \geq 0$ (not assumed to be positive semi-definite). Further, $\mathbf{M} \in \{0,1\}^{n \times n}$ represents the symmetric mutual exclusion (mutex) constraints matrix.

Mutual Exclusion Constraints   Logical binary relations play an important role in context of finding maximum cliques or rather sub-graphs inside a given correspondence graph. They constitute a powerful instrument for modelling incompatible vertex combinations which shall not be part of the same result. In addition to this, they reduce the search space and eliminate invalid alignments beforehand. Thus, enforcing these relations during the actual matching process significantly increases the accuracy and the robustness. It is interesting to know that mutexes are able to express both one-to-one and many-to-one relationships.

Definition 2.4.3. *A correspondence graph $G^\star$ holds the pairwise $(\mathcal{K} \times \mathcal{D})$ similarity between the elements of two given sets, $\mathcal{K}$ and $\mathcal{D}$. Moreover, it models the pairwise similarity between all pairs of $(\mathcal{K} \times \mathcal{D}) \times (\mathcal{K} \times \mathcal{D})$. While the former affiliations are known as unary potentials (vertices weights), the latter are binary relations expressed by the weights on the graph edges (except for reflexive relations).*

Unary Potentials   Vertices weights residing on the diagonal of $\mathbf{A}$, indicating the similarity between two elements $u \in \mathcal{D}$ and $v \in \mathcal{K}$ which shall be matched.

Binary Potentials   Weights assigned to the graph edges and placed as off-diagonal elements inside the affinity matrix. These binary relations describe the pairwise similarity or rather consistency between two unary potentials, respectively. Their significance is additionally strengthened by the mutual exclusions.

In context of the projects using this method, these affiliations are calculated in analogy to the papers [MYL13] and [ML12] used as reference here. Nevertheless, once the affinity matrix is populated by these potentials, the sub-graph with the largest total weight is searched satisfying all given mutex constraints ($\mathbf{M}$). Strictly speaking, just by taking into account the sum of unary and binary relations, the sub-graph can be determined by maximising (2.23) under the conditions modelled in $\mathbf{M}$. Even though this solution might appear easy at a first glance, its implementation is a non-trivial task. Its complexity additionally increases by the fact that this kind of optimisation is known to be NP-hard. In order to tackle this issue, the computational complexity of (2.23) is relaxed by transforming the discrete vertex selection vector x to a continuous one (cf. [ML12]):

$$\underset{x}{\arg\max} \quad g(x) = x^T A x \quad \text{s.t.} \quad x^T M x = 0, \, x \in [0,1]^n \quad . \tag{2.24}$$

This approach is typically used in practice for solving this kind of optimisation problem. Thus, the actual contribution of the authors can be found in a further transformation step, namely the movement of the mutex constraints to the target function (with $\gamma \gg 0$):

$$\underset{x}{\arg\max} \quad f(x) = x^T W x = x^T A x - \gamma x^T M x \quad \text{s.t.} \quad x \in [0,1]^n \quad . \tag{2.25}$$

"[The] key property [of the proposed algorithm] is, that if $\gamma > \max_i \sum_j A_{ij}$ and if the solution $x^\star$ is discrete, then $x^\star$ is guaranteed to satisfy all mutex constraints". [MYL13]

2.4.3 Matching of Time Series

After a deeper discussion about matching concepts based on graph structures in the previous section, the following content is dedicated to the topic of matching time series. Time series can be employed in plenty of various applications, e.g. signal processing, tracking and pattern recognition to mention some of them.

Definition 2.4.4. *The term time series represents a successively ordered collection of observations x_t which have been measured over a finite time interval $t = 1, 2, \ldots, n$. In other words, a time series can be viewed as a linear sequence of data points having a natural order.*

Most often time series are used to model evolutionary processes subjected to dynamically changing variables over time. In context of this work, time series are employed to describe shape variations of geometrical paths by data items calculated as real-valued quantities. Therefore, a fixed number of equidistantly distributed points are placed along the entire path. The idea of this approach follows the assumption that similar paths are similar in their time series. Thus, the determination of the similarity between two paths can be implemented by computing the similarity between two time series. This calculation is finally undertaken by matching algorithms stated above.

Nowadays literature knows plenty of methods to match time series, e.g. the Longest Common Subsequence [WF74; BHR00] or the Maximum Variance Matching [Lat+07a]. Being aware of them, the focus of the following content is set to the Dynamic Time Warping, the Optimal Subsequence Bijection and the Earth Mover's Distance approach since they are frequently used in this thesis. For all subsequent explanations, let $a = (a_1, a_2, \ldots, a_N)$ and $b = (b_1, b_2, \ldots, b_K)$ be two time series, with $\|a\| = N$ and $\|b\| = K$. Furthermore, both series are not assumed to be equal in their lengths, but *discrete* and linearly sampled in time.

2.4.3.1 Dynamic Time Warping

A popular algorithm to find the *optimal* alignment between two time series is known as Dynamic Time Warping (DTW). The algorithm calculates a (pseudo) distance value indicating the dissimilarity between the given sequences a and b. In this context, the term *optimal* refers to the alignment with the minimum distance. However, in order to obtain the *overall* distance, a *local* distance function is required to return the pairwise distance between two elements a_i and b_j, respectively. Since $a_i, b_j \in \mathbb{R}$, a suitable distance function $d(\cdot, \cdot)$ might be defined as:

$$d : \mathbb{R} \times \mathbb{R} \to \mathbb{R}_0^+ \quad . \tag{2.26}$$

Simple instances are the *Euclidean* or the *Manhattan* distance metric (cf. [TK08]). Once the distance measure is known, a cost matrix $\mathbf{C}$ can be calculated carrying the pairwise distance of all elements in a and b. The actual matching is then realised by a non-linear time "warping" in order to determine the overall dissimilarity between both sequences. Strictly speaking, a path in $\mathbf{C}$ is searched with the lowest distance compared to all the other ones. According to [Mül07], a valid warping-path $r^\star$ is defined as follows:

Definition 2.4.5. *A path r composed as a sequence of correspondences $(r_1, r_2, \ldots, r_M)$ with $r_l \in a \times b$ is a valid warping-path $r^\star$ if it fulfils the following conditions:*

- *Boundary condition: $r_1 = (a_1, b_1) \wedge r_M = (a_N, b_K)$*

- *Monotonicity condition: The algorithm is not allowed to go backwards neither in row nor in column of $\mathbf{C}$. This means, the sequence indices (i, j) of two successive path elements $r_l = (i_l, j_l)$ and $r_{l+1} = (i_{l+1}, j_{l+1})$ have to be $i_l \leq i_{l+1}$ and $j_l \leq j_{l+1}$.*

- *Step size condition: The algorithm is not allowed to skip elements in $\mathbf{C}$ neither in row nor in column: $(i_{l+1} - i_l, j_{l+1} - j_l) \in \{(1,0), (0,1), (1,1)\}$ with i and j being two indices.*

Hence, the actual "warping" is realised by a one-to-many matching scheme which corresponds to an implicit *deletion* or *insertion* of certain elements inside the sequences. This enables the algorithm to react on non-linear variations in the time domain. This matching behaviour is implemented by means of the *dynamic programming* (cf. Section 2.2.3). Given a cost matrix $\mathbf{C}$, its cheapest warping-path is found based on a new cost matrix $\mathbf{C}^\circ$ by transforming the original values in $\mathbf{C}$ as illustrated below:

$$\mathbf{C}_{i,j}^\circ = \mathbf{C}_{i,j} + \psi(i,j) \quad , \tag{2.27}$$

with

$$\psi(i,j) = \begin{cases} 0, & \text{if } i = 0 \wedge j = 0 \\ \mathbf{C}_{0,i-1}^\circ, & \text{if } j = 0 \\ \mathbf{C}_{j-1,0}^\circ, & \text{if } i = 0 \\ min\{\mathbf{C}_{0,i-1}^\circ, \mathbf{C}_{j-1,0}^\circ, \mathbf{C}_{j-1,i-1}^\circ\}, & \text{if } i > 0 \wedge j > 0 \end{cases} . \tag{2.28}$$

With access to $\mathbf{C}^\circ$, the alignment with the minimum distance is identified by selecting those rows and columns assigned to the lowest values. This final step is known as *backtracking*. Please remember that this procedure is subjected to the constraints given in Definition 2.4.5. An exemplary alignment is shown in Figure 2.18. By summing up all values which have been selected during the backtracking stage, the overall dissimilarity between both sequences is obtained. In order to retrieve further details as well as information about further variations of the DTW, the reader is referred to [Mül07] and [TK08].

2.4.3.2 Optimal Subsequence Bijection

Another method for matching time series is the so-called Optimal Subsequence Bijection (OSB). The OSB has been proposed by Latecki et al. in [Lat+07b] and can be understood as an extension of the DTW trying to compensate its weaknesses, e.g. that the start and the end point has to be kept for the final result. Like the DTW, the OSB requires two sequences a and b which are not assumed to be of the same length.

> "The goal of OSB is to find subsequences a' of a and b' of b such that a' best matches b'." [Lat+07b]

In other words, the OSB is an elastic time matching which is realised by detecting the shortest path through the given cost matrix similar to the DTW or the Minimum Variance Matching. Moreover, the OSB is able to preserve the order of the sequence and to additionally skip elements in both series with the aim to exclude outliers from the final alignment. Please be aware of the authors' warning that skipping too many elements of a sequence rises the risk of accidental matches. In order to prevent this abnormal behaviour, they introduce a penalty for skipping elements (discussed below). Figure 2.18 illustrates the advantage of using the OSB in comparison to the DTW which is not able to discard outliers.

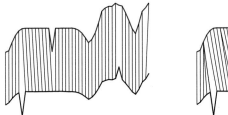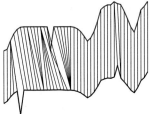

Figure 2.18: The figure shows the alignment result of the OSB in comparison to the DTW and demonstrates the benefit of skipping outliers. (**Left:**) Since the OSB is able to ignore elements in both sequences, a better correspondence configuration can be achieved. (**Right:**) DTW alignment without removing any outliers from the matching (cf. [BL08]).

Even though the originally proposed algorithm provides a higher degree of freedom, it is not going to be discussed in further detail. Instead, the reader can refer to [Lat+07b] while this section switches to a slightly modified version of the OSB which has been introduced in [BL08].

> "[The] goal is to find the best possible correspondence of sequence a to a sub-sequence b' of b." [BL08]

Nevertheless, the objective is still to localise two sub-sequences for the purpose of best matching, but this time the sequence a is privileged. Strictly speaking, the algorithm is looking for the best mapping $b_{f(i)}$ towards the elements in a. Moreover, the mapping function $f(\cdot)$ (restricted to the indices $i_k \in \{1,\dots,N\}$) is designed to be a *bijection*. This is done by introducing an additional node enabling a many-to-one mapping for all i_k needed to be skipped (cf. [BL08]). The starting point for both versions is the creation of the cost matrix **C** based on a local distance function:

$$d : \mathbb{R} \times \mathbb{R} \to \mathbb{R} \quad . \tag{2.29}$$

By means of this function, the pairwise distances between all elements of sequence a and b are calculated. The authors confirm that there are no restrictions on the distance function.

Thus, one can freely choose which one to take. By selecting an appropriate instance for $d(\cdot,\cdot)$, the cost matrix $\mathbf{C}$ can easily be populated. The penalty for skipping elements, referred to as *jumpcost* in the following, is approximated in form of a constant value:

$$jumpcost = \text{mean}(\min_i(\min_j(d(a_i, b_j))) + \text{std}(\min_i(\min_j(d(a_i, b_j)))) \quad . \tag{2.30}$$

Here the original notation has been used to formulate the proposed two-tier calculation scheme. First, the closest element b_j is searched for every a_i. Second, the *mean* and the *standard deviation* are generated based on these minimum distances. The *jumpcost* is then retrieved by adding both values. Please recognise that these minimum distances can already be gathered during the population of $\mathbf{C}$.

The actual alignment problem is finally considered as finding the shortest path in a directed acyclic graph G°. For this purpose, $\mathbf{C}$ has to be transformed to G°, so that each matrix element represents one single vertex. The more challenging task is to define the weights on the edges of G°. This is realised with the following binary relation:

$$\varrho((i,j),(i',j')) = \begin{cases} d(a_i, b_j), & \text{if } i+1 = i' \wedge j+1 \leq j' \\ (i'-i-1) \cdot jumpcost, & \text{if } i+1 < i' \wedge j+1 \leq j' \\ \infty, & \text{otherwise} \end{cases} \quad . \tag{2.31}$$

A closer inspection of these conditions reveals that the algorithm is not allowed to go backwards neither in row nor in column. Moreover, the skipping of columns (elements in b) is not explicitly penalised while it is the case for the rows (members of a).

"The intuition is that we expect all elements of sequence a to find a partner element in sequence b with possibly skipping some elements of b." [BL08]

However, if the lengths of both sequences are equal or if sequence a encompasses more elements than b, a skipping of values in b would lead to an implicit penalisation of a since each alignment f is an injection on a. Thus, an exclusion of elements in b would additionally imply the skipping of members in a and this, in turn, is penalised by the proposed algorithm. Please further recognise that the condition $\mathbf{i+1} = \mathbf{i'} \wedge j+1 \leq j'$ only holds for the successive row $i+1$ in $\mathbf{C}$.

Finally, the *optimal* alignment is found by a shortest path algorithm on G°, while the total distance for two sequences can be calculated according to:

$$d(a,b,f) = \frac{1}{N} \sum_{i=1}^{N} d(a_i, b_{f(i)})^2 \quad . \tag{2.32}$$

2.4.3.3 Earth Mover's Distance

Like the methods above, the Earth Mover's Distance (EMD) constitutes a powerful instrument to determine the distance between two time series. The reason why it is discussed in more detail is attributed to its property of being able to consider a further dimension. In contrast to the other approaches, the EMD takes two weight sequences $\hat{a}$ and $\hat{b}$ as additional input to the two time series a and b. The notion behind the concept of the EMD is the reformulation of the problem (the distance calculation) to the well-known transportation problem. Thus, finding a solution to that problem does also retrieve the desired distance.

> "Intuitively, given two distributions, one can be seen as a mass of earth properly spread in space, the other as a collection of holes in that same space. Then, the EMD measures the least amount of work needed to fill the holes with earth." [RTG00]

Strictly speaking, the two sequences a and b determine the position of the piles or holes while their weights or rather volumes ($\hat{a}$ and $\hat{b}$) represent the possible supply or demand of them, respectively. The so-called *ground distance*, a local distance measure, calculates the pairwise distance between the points in a and b. The ground distance function is defined as:

$$d : \mathbb{R}^n \times \mathbb{R}^n \to \mathbb{R}_0^+ \quad . \tag{2.33}$$

Thus, in context of the transportation problem, the ground distance defines the displacement between the piles and holes. Now the aim is to minimise the transportation costs needed to fill up the holes. In the following, the time series and their weight sequences are coupled to signatures: $A = \{(a_1, \hat{a}_1), \ldots, (a_N, \hat{a}_N)\}$ and $B = \{(b_1, \hat{b}_1), \ldots, (b_K, \hat{b}_K)\}$. The actual transportation costs are calculated as follows:

$$c(A, B, \mathbf{F}) = \sum_{i=1}^{N} \sum_{j=1}^{K} d(a_i, b_j) \cdot \mathbf{F}_{i,j} \quad , \tag{2.34}$$

where the flow matrix $\mathbf{F}$ determines the amount of earth which has to be transported from the pile a_i to the hole b_j. Thus, the concrete task is to find a flow $\mathbf{F}$ that minimises the overall costs $c(\cdot, \cdot, \cdot)$ (cf. [RTG00]). This minimisation is then performed under the constraints:

$$F_{i,j} \geq 0, \quad 1 \leq i \leq N, 1 \leq j \leq K \quad , \tag{2.35}$$

$$\sum_{j=1}^{K} F_{i,j} \leq \hat{a}_i, \quad 1 \leq i \leq N \quad , \tag{2.36}$$

$$\sum_{i=1}^{N} F_{i,j} \leq \hat{b}_j, \quad 1 \leq j \leq K \quad , \tag{2.37}$$

$$\sum_{i=1}^{N} \sum_{j=1}^{K} F_{i,j} = \min\left(\sum_{i=1}^{N} \hat{a}_i, \sum_{j=1}^{K} \hat{b}_j\right) \quad . \tag{2.38}$$

The conditions are straightforward but shall briefly be addressed for the sake of completeness. The first condition ensures that the earth is entirely transported in one direction, condition two and three restrict the method of carrying only the actual available volume. In other words, it prevents that the transported earth exceeds the volume of a pile or hole. The last constraint guarantees that the maximum amount of earth (the weight of the lighter distribution) is going to be transferred by the algorithm. Finally, the minimum flow can be calculated by an appropriate method for solving this optimisation problem subjected to the constraints above. The most famous technique in this area is the simplex algorithm. Further information about *linear programming* as well as the simplex method is provided in Section 2.2.4. Moreover, the work of David Gale [Gal07] offers a good introduction to this topic. Finally, the optimisation result has to be normalised in order to stay comparable to time series with different lengths and to avoid favouring smaller signatures in context of partial matchings (cf. [RTG00; Sch05]:

$$c^{(EMD)}(A,B) = \frac{\underset{F}{\operatorname{argmin}} \; \left(\sum_{i=1}^{N} \sum_{j=1}^{K} d(a_i, b_j) \cdot F_{i,j}\right)}{\sum_{i=1}^{N} \sum_{j=1}^{K} F_{i,j}} \quad . \tag{2.39}$$

Chapter 3

2D Skeleton Graph Matching for 2D Object Retrieval

This chapter introduces the so-called Path Similarity Skeleton Graph Matching (PSSGM) [BL08] proposed by the authors Bai and Latecki. Their work presents a skeleton-driven graph matching approach for the purpose of 2D object retrieval. Moreover, their method constitutes the foundation of the underlying work and thus, it is going to be thoroughly explained in this chapter. The discussion involves the skeleton extraction, the concept of the shortest path, the feature generation and the actual matching process.

Being aware that the method significantly affects the research of the underlying work, it has been re-implemented in [Hed+13] with the goal to analyse it in more detail as well as to verify its promising results which had been presented in the original paper [BL08]. The findings of this investigation are then discussed in Section 3.3. This section clearly figures out the limitations and the strengths of the PSSGM and hence, important knowledge for all subsequent projects is obtained which needs to be considered during the mapping of that technique into new context areas.

3.1 Fundamental Concepts

3.1.1 Discrete Curve Evolution

The Discrete Curve Evolution (DCE) was introduced in 1999 by the authors Latecki and Lakämper in [LL99a]. The DCE belongs to the class of scale-space evolution methods capable of partitioning an arbitrary object shape into a set of 2D curve segments. The intention behind this is to simplify the object's geometry by emphasising the most significant parts of the given shape. The degree of abstraction can be steered by a parameter. The core idea of DCE relies on the assumption that every digital boundary curve of an object Ω can be represented as a finite polygon $\mathcal{P}$ without any loss of information.

> "In opposite to standard approaches in scale-spaces, [this] evolution is guided neither by differential equations nor Gaussian smoothing, and it is not a discrete version of an evolution by differential equations" [LL99b].

In more detail, the DCE returns a subset $(\partial\Omega)^c$ of the object boundary points $\partial\Omega$ that mostly preserve the appearance of its shape. Furthermore, $(\partial\Omega)^c$ is smoothed in terms of alleviating segmentation noise and positional errors as discussed later in this section. The actual simplification of the object's boundary is iteratively performed by removing (irrelevant) points from $\partial\Omega$, while keeping the most structurally relevant shape information. Figure 3.1 shows three differently evolved stages of a boundary forming the shape of a fish[1] after applying the DCE to it.

Figure 3.1: The figure demonstrates the evolutionary character of the method by performing three successive steps of the DCE applied to the instance of a fish shape. The selected points in time clearly illustrate the simplification of the boundary by removing irrelevant points.

In the following, the implementation of the DCE shall briefly be introduced. Therefore, let a be single line and $\mathcal{P}^{(0)}$ the contour $(\partial\Omega)$ of an object (Ω) represented as a polygon that contains M vertices. In each iteration t, the polygon is reduced by a single vertex $\mathbf{p}_{i=1,...,M}$ having the lowest contribution to the overall shape appearance. After removing $\mathbf{p}_i$ from $\mathcal{P}^t$, its predecessor $\mathbf{p}_{i-1}$ is connected with its successor $\mathbf{p}_{i+1}$ in order to create a new (polygon) line $\overline{\mathbf{p}_{i+1}\mathbf{p}_{i+1}}$ which bridges the gap at $\mathbf{p}_i$. The decision whether a point has to be removed or not is controlled by a *relevance measure* $\psi : \mathbb{R}^2 \rightarrow \mathbb{R}_0^+$ indicating the contribution of $\mathbf{p}_i \in \mathcal{P}$ to the object's shape (cf. [LL99a]):

$$\psi(\mathbf{p}_i) = \frac{g(a_1,a_2)l(a_1)l(a_2)}{l(a_1)+l(a_2)} \quad , \tag{3.1}$$

where $l(\cdot)$ returns the normalised length of an arbitrary (polygon) line segment and $g(\cdot,\cdot)$ determines the angle at $\mathbf{p}_i \in \mathcal{P}^t$ with $a_1 = \mathbf{p}_{i-1} - \mathbf{p}_i$ and $a_2 = \mathbf{p}_{i+1} - \mathbf{p}_i$. The actual normalisation

---

[1]www.dabi.temple.edu/~shape/shape/intro_applet1.html, [online: 19th August 2015]

is performed by employing the *total* number of vertices in $\mathcal{P}$ at time t. The higher the value of (3.1), the higher the contribution of $\mathbf{p}_i$ to the object's appearance and thus, it is kept in the current evolution step t. In consequence to this, only points with low relevance are deleted from the polygon and are not further considered during subsequent tasks. The result of Equation (3.1) is similar to a cost value for linearising an arc between two vertices. At this point, the reader is referred to [LL99a] to obtain more detailed information about the relevance measure, e.g. its correlation to the tangent space.

Summarised, the DCE can be viewed as a hierarchical decomposition of the initial contour polygon $\mathcal{P}^{(0)}$ into a set of vertex partitions with different numbers of elements. The further the polygon evolves, the lower is its amount of vertices: $\|\mathcal{P}^t\| > \|\mathcal{P}^{t+1}\|$. In contrast to a diffusion process which only operates based on translation vectors, the DCE takes global contour information into account with the result of being a more accurate alternative. In addition to this, the application of DCE eliminates distortions caused by noise and displacement errors which might occur during the process of extracting of the object's contour (e.g. image segmentation). These properties as well as the simplified shape by itself are highly beneficial in context of skeletonisation as introduced in Chapter 2. Please recall that the method first eliminates those vertices with the smallest shape contribution by emphasising the structurally significant parts of the shape (cf. [LL99a; LL99b; LL00]). Strictly speaking, by defining the skeleton end points as those contour vertices having a relevance value above a certain threshold, a robustly performing skeleton growing technique can be presented in [BLL07]. This approach is also used as pre-processing stage in [BL08] joining together the process of extracting the skeletal structure with that of the skeleton pruning.

3.2 Path Similarity Skeleton Graph Matching

In [BL08] a skeleton graph matching approach is presented to find best fitting correspondences between two 2D objects by exploiting their skeletons. Therefore, a novel features set is proposed which is straightly derived from the skeletal structure. Concerning the task of skeletonisation, the authors employ the Discrete Curve Evolution (DCE) as described in Section 3.1.1. The DCE is highly suitable for this purpose: It preserves the most significant parts of the object, whereas spurious branches are suppressed. As a result of this, the quality of the PSSGM input can be increased drastically. The actual matching is then performed on the skeleton end points and the concept of emanating shortest paths. By taking these paths, a so-called path distance or rather dissimilarity is calculated which is subsequently used to establish the desired correspondences. Therefore, the paths are converted to sequences of radii according to the maximal disks which have been fitted into the boundary as demonstrated in Figure 3.3. Having this data, matching costs are computed by taking into account all shortest paths emanating from a certain pair of end points. By arranging these values in form of a matrix and by passing them to the OSB function (cf. Section 2.4.3.2), similarity indicators are obtained for each point pair between both skeletons. Finally, the overall shape distance is calculated by executing the Hungarian method (Section 2.4.2.1).

3.2.1   2D Object Representation by Skeleton Graphs

The object representation is crucial for the success of the matching process. With respect to the objects' characteristics, the representation type has to be carefully selected to prevent the loss of distinctive properties. In context of the PSSGM, the skeletal structure is first extracted from the object and subsequently used to describe the most significant regions of the shape. Section 2.4.1 introduces several techniques to compress an object to its skeleton. Here the DCE is incorporated to undertake this job. As already discussed in Section 3.1.1, the use of the Discrete Curve Evolution has attractive properties towards the skeleton extraction, e.g. the implicitly performed skeleton pruning. Afterwards, two skeleton graphs G and G' are going to be constructed which interpret all end points as a set of vertices. Having these graphs, the idea is to match these nodes ($v_i \in G, v'_j \in G'$) between both skeletons. Therefore, the shortest paths (or geodesics) are determined on the skeleton for all edges in G and G' as illustrated in Figure 3.2. In the following content, a geodesic is retrieved by the function $\rho(\cdot, \cdot)$ capable of taking two vertices as input. Afterwards, it returns the smallest set of skeleton points connecting these given vertices. The actual path detection can easily be implemented

Figure 3.2: The figure shows the skeleton of a bird and two exemplary geodesics which have been derived from this structure. Please notice that there are more than two paths which have to be established if all end point are taken into consideration (cf. [Hed+13]).

by exploiting a shortest path algorithm, e.g. the Dijkstra method (cf. Section 2.2.5). Once all shortest paths are available, they are respectively converted to a sequence of numbers describing the path uniquely. Therefore, the authors of [BL08] propose a sampling scheme which incorporates the object's contour. Strictly speaking, by distributing a fixed number K of equidistantly aligned points along each path, sample positions are defined which are used to calculate a maximum inscribed disk that touches the contour at, at least, two or more places. The principle is also illustrated in Figure 3.3, where the sample locations are depicted as red pixels and the disks as yellow circles.

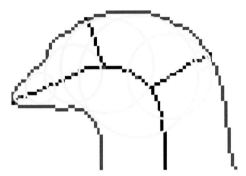

Figure 3.3: The figure illustrates the path sampling principle introduced in [BL08]. Operating on the skeleton shown in Figure 3.2, the concept of using maximum inscribed disks for the purpose of representing the shortest paths is demonstrated. The red points distributed along the path (drawn in black) are the sample locations $\mathbf{q}_i$, whereas the corresponding maximum disks are drawn as yellow circles (cf. [Hed+13]).

Since the centre of the disk is already known by the sample position $\mathbf{q}_{m,i=1,\ldots,K}$ on the skeleton path m, the distance transform $D(\cdot)$ which has already been used during the skeletonisation, can be recruited in order to receive the radius $r_{m,i}$ of the maximum disk at this location. The actual value for r is then approximated by taking the return value of $D(\mathbf{q}_i)$ (at the sample location $\mathbf{q}_i$) coupled with a denominator responsible for being scale invariant during the matching:

$$r_{m,i} = \frac{D(\mathbf{q}_i)}{\frac{1}{N}\sum_{j=1}^{N} D(\mathbf{p}_j)} \quad . \tag{3.2}$$

Please observe that the value of the denominator in (3.2) correlates to the object's size, where $\mathbf{q}_j$ iterates over all N pixels which are accommodated inside the object Ω. In more detail, the denominator calculates the average distance to the object's boundary $\partial\Omega$. By gathering radii, the final feature vector is obtained by concatenating all values in form of a sequence or rather K-dimensional feature vector $\mathbf{x}$:

$$\mathbf{x}_m = (r_1, r_2, \ldots, r_i \ldots, r_K)_m^{\mathsf{T}} \quad . \tag{3.3}$$

3.2.2   Path Distance as Dissimilarity Measure between Points

Definition 3.2.1. *The path distance d indicates the dissimilarity between two shortest paths $\rho \subset S$ and $\rho' \subset S'$ being part of the skeletons S and S' which, in turn, are representing the objects Ω and Ω', respectively (cf. [BL08]).*

The path distance is a crucial factor regarding the success of the matching process. The path distance is used to determine the dissimilarity between two sampled paths. By taking the path distances of all paths which are respectively emanating from $v \in G$ and $v' \in G'$, the matching costs for v' and v are calculated. In the following, the path distance is computed between two exemplary shortest paths $\rho(v,u) \in G$ and $\rho(v',u') \in G'$:

$$d(\rho(v,u),\rho(v',u')) = \sum_{i=1}^{K} \frac{(r_i - r_i')^2}{r_i + r_i'} + \eta \frac{(l(\rho(v,u)) - l(\rho(v',u')))^2}{l(\rho(v,u)) + l(\rho(v',u'))} \quad , \qquad (3.4)$$

with $l(\cdot)$ being a function that returns the length of a given path. The incorporation of the path lengths is reasonable since the proposed sample scheme is not able to reflect them adequately (cf. Equation (3.2)). In addition to this, the coefficient $\eta \in \mathbb{R}^+$ controls the influence of this component on the overall distance value in form of a weight factor. It is worth mentioning that the authors do not penalise any path deformation in (3.4). They justify this step with the aim to improve the recognition performance of articulated shapes, e.g. animals and humans which are typically subjected to strong deformations [BL08].

3.2.3   2D Object Matching Using Skeleton Graphs

This section discusses the matching process responsible for establishing correspondences between two skeletons based on their skeleton graphs. Therefore, let G and G' be two graphs representing two skeletons consisting of $\overline{K} + 1$ and $\overline{N} + 1$ end nodes, respectively. Moreover, it is assumed that $\overline{K} \leq \overline{N}$. According to [BL08], the vertices of G and G' are denoted by $v_{i=1,...,\overline{K}}$ and $v'_{j=1,...,\overline{N}}$. The notion behind this procedure is to exploit the *path distance* as defined in Equation (3.4) to achieve an optimal alignment of the skeleton nodes.

For this purpose, all possible pairs of vertices between G and G' are going to be examined under the assumption that almost identical paths are expected if two nodes indicate the same shape area in both objects. Hence, the matching costs $c(v_i, v'_j)$ are calculated based on all skeleton paths emanating from v_i and v'_j. Therefore, the end nodes are arranged sequentially by traversing the contour clockwise starting at v_i and v'_j, respectively. According to the notation used in [BL08], the current observation is depicted by v_{i0} and v'_{j0}. The order of the remaining points is then determined by their position during the traversal of the contour in clockwise direction: $[v_{i0}, v_{i1}, ..., v_{i\overline{K}}]$ and $[v'_{j0}, v'_{j1}, ..., v'_{j\overline{N}}]$. Pursuant to this

order, the path distance values are calculated for all paths emanating from v_{i0} and u_{j0}:
$[(v_{i_0}, v_{i_1}), (v_{i_0}, v_{i_2}), \ldots, (v_{i_0}, v_{i\overline{K}})]$ and $[(v'_{j_0}, v'_{j_1}), (v'_{j_0}, v'_{j_2}), \ldots, (v'_{j_0}, v'_{j\overline{N}})]$ in order to arrange them in one cost matrix:

$$
\mathbf{C}^{(OSB)}(v_i, v'_j) = \begin{pmatrix} d(\rho(v_{i_0}, v_{i_1}), \rho(v'_{j_0}, v'_{j_1})) & \cdots & d(\rho(v_{i_0}, v_{i_1}), \rho(v'_{j_0}, v'_{j\overline{N}})) \\ d(\rho(v_{i_0}, v_{i_2}), \rho(v'_{j_0}, v'_{j_1})) & \cdots & d(\rho(v_{i_0}, v_{i_2}), \rho(v'_{j_0}, v'_{j\overline{N}})) \\ \vdots & \ddots & \vdots \\ d(\rho(v_{i_0}, v_{i\overline{K}}), \rho(v'_{j_0}, v'_{j_1})) & \cdots & d(\rho(v_{i_0}, v_{i\overline{K}}), \rho(v'_{j_0}, v'_{j\overline{N}})) \end{pmatrix} . \tag{3.5}
$$

This matrix is then passed to the Optimal Subsequence Bijection (OSB) function as introduced in Section 2.4.3.2 capable of compressing it to one scalar that indicates the matching cost for v_i and v'_j. In order words, this value has to be spent in order to establish a correspondence between these end nodes:

$$
c(v_i, v'_j) = OSB(\mathbf{C}^{(OSB)}(v_i, v'_j)) \quad . \tag{3.6}
$$

In context of PSSGM, the skipping of elements in $\mathbf{C}^{(OSB)}$ by the OSB excludes those paths or rather end nodes which either do not exist in the other skeleton or simply increase the overall matching costs. As penalty or *jump cost* for skipping certain elements in $\mathbf{C}^{(OSB)}$, the same value is used as originally proposed. Using this approximation, the authors of [BL08] expect to obtain an adaptive value assumed to be large enough to detect outliers while being small enough to prevent the skipping of too many elements. Please consult Section 2.4.3.2 to obtain more detailed information.

By collecting the matching costs of all possible arrangements of skeleton end points, the global configuration of node correspondences can be calculated by utilising the Hungarian method (cf. Section 2.4.2.1). Therefore, a further cost matrix $\mathbf{C}^{(G,G')}$ is generated and passed as input to the Hungarian method:

$$
\mathbf{C}^{(G,G')} = \begin{pmatrix} c(v_0, v'_0) & c(v_0, v'_1) & \cdots & c(v_0, v'_{\overline{N}}) \\ c(v_1, v'_0) & c(v_1, v'_1) & \cdots & c(v_1, v'_{\overline{N}}) \\ \vdots & \vdots & \ddots & \vdots \\ c(v_{\overline{K}}, v'_0) & c(v_{\overline{K}}, v'_1) & \cdots & c(v_{\overline{K}}, v'_{\overline{N}}) \end{pmatrix} . \tag{3.7}
$$

The Hungarian method returns the optimal alignment in terms of minimum matching costs. The total dissimilarity between G and G' is finally obtained by summing up all cost values being part of this configuration. A brief introduction of the working principle behind the Hungarian method is given in Section 2.4.2.1 and shall be referred to at this point. However, please bear in mind that this optimisation problem is mapped to a classic assignment problem based on a bipartite graph.

However, due to the fact that G and G' might be unbalanced, dummy nodes have to be incorporated. In [BL08], the authors suggest to add additional rows to (3.7) to enforce the obligatory symmetry property of $\mathbf{C}^{(G,G')}$. In order to populate these new row entries, a constant value is taken defined as the average of $\mathbf{C}^{(G,G')}$. By involving these dummy nodes a "globally consistent one-to-one assignment of all end nodes" is achieved. Strictly speaking, by introducing these hidden nodes, the algorithm is enabled to align even those nodes which do not have a proper matching partner. As already stated above, the total dissimilarity is still obtained as sum of the costs which are part of the final configuration. Please keep in mind that the Hungarian method does not guarantee to preserve any order of nodes and thus, the method runs the inherent risk of generating displaced correspondences. However, according to Bai et al. this potential risk of wrong assignments cannot affect the final score. This conclusion relies on the issue that assignments are established based on their similarity which, in turn, is calculated with respect to the other (adjacent) end nodes. Hence, only symmetry changes can lead to such a disorder in theory.

3.3   Revision of the Path Similarity Skeleton Graph Matching

After introducing the theoretical concept of the Path Similarity Skeleton Graph Matching (PSSGM) approach, this section further evaluates it in terms of robustness and accuracy. This means the study focuses on the methodological perspective with the goal to figure out the method's strengths and restrictions. However, the investigations are primarily intended to discover weak points of the method which require a higher amount of attention. This awareness of potential problems is highly important, especially if the method shall be mapped to other areas, e.g. 3D. Moreover, a deep understanding also supports and simplifies any kind of modifications necessary to adapt the method to new applications. This knowledge in general offers space for improvements and further research.

Recalling that the method is used to calculate the similarity between two shapes based on the best fitting alignment by utilising the objects' skeletons, it had to be re-implemented in order to perform a thorough analysis as presented by Hedrich et al. in [Hed+13].

3.3.1   Strengths

The task of robustly matching objects is challenging in all areas which are dedicated to the subject of object recognition. The problem is even exacerbated in presence of deformation.

Being aware of this situation, Bai et al. addressed the issue of deformable shapes in their work [BL08] based on the idea of representing an object by its skeletal structure. This is absolutely reasonable since skeleton branches typically occur in significant areas of the shape which drastically contribute to the object's perceptual appearance. Thus, the method qualifies itself to be an appropriate instrument to express shape deformations. Figure 3.4 demonstrates this capability of emphasising significant object parts even in case of articulation. Finally, the PSSGM embodies a highly generic character which easily allows its

Figure 3.4: The figure shows the instance of a crocodile. Please recognise that the skeleton branches are accommodated inside the most significant areas of the shape, which are the tail, the legs and the mouth. Especially the crocodile's tail is clearly represented by the skeletal structure even in case of articulation (cf. [Hed+13]).

mapping to further research areas. In other words, the approach is not limited to a specific object class, operating space (2D, 3D) or structure towards the input data (pixel/voxel grid, point cloud). As long as the centre line of an object is available, the method can be used to perform object categorisation tasks. The success of entering new domains can additionally be increased by replacing the proposed feature set with more meaningful information while keeping the present sampling scheme.

Some of these adaptations are going to be discussed later in this thesis. Chapter 5 uses e.g. 3D curves for the purpose of representing the boundary of a 3D object. Even though these line fragments do not natively correspond to the definition of a skeleton, it can be interpreted as a special type of it which enables the application of the PSSGM. In contrast to this, Chapter 6 employs a strongly modified version of the PSSGM to register pre- and post-operative abdominal aorta structures by exploiting their centre lines. Therefore, a more sophisticated feature set and a novel matching technique is employed. However, behind all of these adaptations, the approach stays the same as originally proposed.

3.3.2 Restrictions & Weaknesses

More critical in context of mapping an existing approach to a new application area are latent limitations which might sabotage this intent. Being aware of this, the following content is primarily dedicated to the restrictions imposed by the PSSGM. This investigation is realised by an empirical analysis which has been performed additionally to the analytic one based on the re-implemented code of this method. Moreover, the study aims at methodological problems which typically occur without being influenced by a given configuration. Thus, these problems cannot be tackled by only optimising the underlying set of parameters. Consequently, these issues need to receive more attention during further processing steps in order to prevent an unexpected behaviour in context of the new application.

The problems which are going to be introduced in the following have been observed in connection to (i) *flipped objects*, (ii) *spurious skeleton branches* and (iii) the enforcement of *one-to-one matchings*.

Flipped Objects   Given two objects which belong to the same class, e.g. two elephants, and one of the objects is mirrored at its vertical axis. Assuming this situation, the following observation has been made: The PSSGM is not able to match these two objects appropriately. The problem is raised by the implementation of the OSB which calculates the costs falsely even if the two assignments would establish a correct correspondence. As already stated in Section 2.4.3, the OSB expects the cheapest alignment on the diagonal of the cost matrix $C^{(OSB)}$. In other words, the optimal configuration is searched from the upper left to the lower right corner inside the matrix. This is a valid behaviour if the objects are not mirrored against each other. The explanation for this lies in the working principle of the OSB which does not allow to go backwards neither in rows nor in columns. However, if one of the objects is mirrored vertically, this operating instruction turns out to be the problem which interferes the detection of the optimal alignment with minimum costs. Deeply inspecting the cost matrix which is generated in presence of mirrored objects discovers that the best fitting alignment is now crossing the matrix elements from the upper right to the lower left in $C^{(OSB)}$ as illustrated in Figure 3.5.

Consequently, the OSB does not perform robustly if the data set has not been pre-processed, e.g. with the intent to align the objects in a pre-defined pose. Another solution is given by the authors of [Hed+13] who propose to execute the OSB two times consecutively. While the original setting is used during the first run, the second cycle processes the opposite direction of the matrix. Finally, the similarity of both objects is expressed by the lower cost value. Figure 3.6 demonstrates the application of this workaround based on two exemplary situations. By comparing both results, its improvement becomes obvious. Although it

$$\begin{bmatrix} \cdots & 25 & 16 & 37 & \boxed{0} \\ \cdots & 8 & 11 & \boxed{0} & 37 \\ \cdots & 16 & \boxed{0} & 11 & 16 \\ \cdots & \boxed{0} & 16 & 8 & 25 \\ \cdot^{\cdot^{\cdot}} & \vdots & \vdots & \vdots & \vdots \end{bmatrix}$$

Figure 3.5: The figure illustrates a constructed example outlining a part of the cost matrix as it would appear if the two objects were mirrored against each other. The location of the best fitting correspondence configuration can now be discovered by following the diagonal from the upper right to the lower left corner inside the matrix (cf. [Hed+13]).

performs promisingly, the code adaptation cannot guarantee the matching success in any case. This mirroring issue does also constitute a valid problem in 3D and requires more attention during the mapping process.

Spurious Skeleton Branches A more obvious problem in context of using skeletons are artefacts in their structures. As already discussed in Section 2.4.1, the skeletonisation of an object constitutes a challenging task which typically suffers from the generation of irrelevant curves. The deletion of these *spurious skeleton branches* requires a post-processing step, namely the skeleton pruning.

In [BL08], the skeleton pruning is performed implicitly by utilising the DCE introduced in Section 3.1.1. Although the method is capable of producing stable skeleton end points residing on the object's shape, missing or occluded contour regions (as it might occur in presence of strong articulations) pushing the method to its limit. These deformations rise the risk of irrelevant skeleton branches which, in turn, affect the matching performance of the PSSGM. Thus, the technique highly relies on the stability and unity of the skeletal structure towards instances of the same object class. Moreover, this meaningless data confuses the algorithm in terms of selecting appropriate matching partners. Figure 3.7 demonstrates the power of these interfering skeleton branches. By taking a closer look at this example, one can discover multiple differences between both skeletal structures responsible for this disastrous result. Please bear in mind that this issue is further stressed by the enforcement to assign every skeleton node in the graph.

The PSSGM tries to tackle this problem by incorporating dummy nodes acting as latent matching partners for these disturbing elements. However, the calculation of an adequate dummy weight is difficult and a wrong value might have unexpected side-effects. It is worth mentioning that even in cases where the weight is chosen adequately, the negative impact of

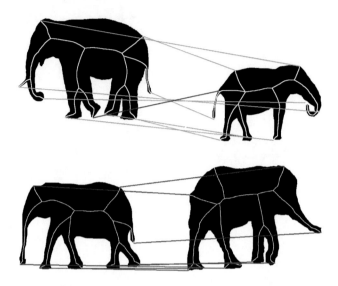

Figure 3.6: The figure demonstrates the application of the workaround which executes the OSB two times instead of only once. While the matching behaves mostly as expected in the upper row, it fails entirely in the lower one. The problem is ascribed to the implementation of the OSB which prohibits to go backwards neither in rows nor in columns. Please notice that the wrong alignment in the upper case is the result of the missing elephant leg (cf. [Hed+13]).

spurious branches cannot be fully suppressed during the alignment procedure. The severity of these structural anomalies has additionally been verified by removing these branches manually with the result of a drastically increased matching quality (cf. Figure 3.7).

The Enforcement of One-to-One Matchings   A further point of interest correlating with the topic of spurious branches is the enforcement of *one-to-one correspondences*. This restriction is imposed by the Hungarian method and needs to be considered during further research activities. In more detail, the establishment of one-to-one alignments does always lead to trouble if two matching sets are encompassing a different number of elements. Hence, not every element in the first set can be assigned to a compatible partner in the second one. In addition to this, the Hungarian method is not capable of skipping certain nodes in order to exclude outliers from the final configuration. Considering the concrete case of the PSSGM, the authors of [BL08] propose the incorporation of *dummy nodes* with the intent to realise the

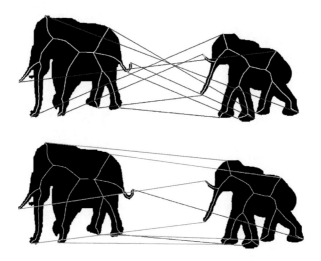

Figure 3.7: The figure illustrates the worst case of a possible matching result which has been affected by irrelevant skeleton branches. Closely inspecting the upper row of the figure, multiple differences can be discovered between both skeletons residing inside the tail and the leg. By removing the spurious branch in the tail, the result can be improved remarkably. However, due to the one-to-one matching condition there are still inevitable misalignments in the lower row (cf. [Hed+13]).

matching of two sets differing in their amount of vertices. Strictly speaking, by involving such dummy nodes the shape of the cost matrix is arranged to be symmetric.

Figure 3.8 illustrates this problem based on the following situation: Let Ω and Ω' be two objects, the query and the target, which contain the same number of end points: $v_{i=1,...,K} \in G$ and $u_{j=1,...,N} \in G'$ with $K = N$. Moreover, assume that one node in G does not correspond to another one in G' and vice versa. Closely inspecting this constructed scene, it becomes obvious that these graphs cannot be aligned correctly based on the concept of determining one-to-one correspondence. Instead, an optimal solution would reject these incompatible points from the final matching. Such an exclusion of nodes can be achieved, e.g. by the assignment strategy of the MWC approach capable of processing only a subset of the given nodes (cf. Section 2.4.2.2).

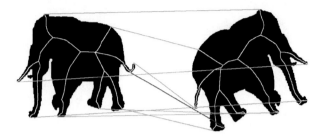

Figure 3.8: The figure shows a further example where the one-to-one alignment led to the occurrence of a falsely generated correspondence pair. The problem arose due to the fact that both shapes contain one incompatible node which has no appropriate matching partner in the other skeleton graph and vice versa (cf. [Hed+13]).

3.4   Summary

This section concludes the discussion started in the previous sections devoted to the Path Similarity Skeleton Graph Matching (PSSGM) approach. This method, presented by Bai et al. in [BL08], aims at the matching of 2D shapes based on their skeletons. Therefore, the object's centre line is first extracted from both the query and the target object. For this purpose, the Discrete Curve Evolution (DCE) is employed capable of simplifying a certain boundary to a small set of meaningful contour points. Strictly speaking, the DCE by itself does not produce the actual skeletal structure but it strongly contributes to this task by determining its end points. Therefore, the most significant perceptual regions of the shape are emphasised by removing irrelevant contour points. The procedure is repeated until a significant set of shape points has been retrieved. By considering these contour locations as end points of the skeleton, they can be exploited as seeds for a skeleton growing approach which straightly connects to the DCE. Hence, in comparison to other state-of-the-art skeletonisation techniques, the skeleton is not obtained by shrinking or rather contracting the shape, but by performing an iteratively operating growing mechanism starting at those points returned by the DCE.

Once the skeletons have been generated, the concept of shortest paths is introduced with the intent to set these paths into relation by calculating their dissimilarity based on the so-called path distance. In other words, each path is sampled by a fixed number of points defining the centres of maximum disks which are determined at these locations. By gathering the radii of all disks, a sequence of values is obtained, the feature vector. Taking the sequences of two paths as well as their path lengths, the path distance can be generated.

This dissimilarity value is then calculated for all shortest paths emanating from a pair of end nodes, e.g. $v \in G$ and $v' \in G'$. Therefore, all remaining end points of the skeleton are ordered clockwise to the object's contour. The actual matching costs for (v, v') are finally obtained by passing all path distances (in form of a time series) to the Optimal Subsequence Bijection (OSB) which determines the optimal alignment with minimum costs. This is repeated for all pairs of correspondences that can be established between the query and the target object. Please notice that the OSB is able to skip outliers while it preserves the order of the time series. If all combinations of skeleton end nodes are processed, the final configuration of correspondences is determined by exploiting the Hungarian method.

The original results presented in [BL08] attest the PSSGM a highly robust and accurate matching performance in regard to a broad variety of objects. Even in presence of deformations or occlusions, the method performs well. However, these results could only partially be reconstructed by the re-implemented version of the PSSGM which has been employed in this project. This might have several reasons, on the one hand, the originally used parameter set has not been available for the internal evaluation and thus, it cannot be guaranteed that the same configuration has been taken. On the other hand, it is not known how the problem of spurious skeleton branches has been tackled in order to prevent misclassification. Especially these irrelevant structure elements strongly affect the quality of the retrieval performance. Furthermore, the implementation of the OSB runs the potential risk of corrupting the PSSGM outcome if object contours are flipped to each other.

Summarised, the skeletal structures have to be perfectly shaped in order to obtain an optimal matching result. These structural anomalies have been detected as the biggest issue in context of the PSSGM. Neglecting the treatment of these outliers has a drastic impact on the matching performance of this method. Being aware of these problems, subsequent chapters are able to react appropriately to these obstacles in order to design robustly operating recognition systems by reusing the entire PSSGM approach or only parts of it. For example, Chapter 5 substitutes the Hungarian method with the approach of Maximum Weight Cliques capable of determining partial matching configurations.

Chapter 4

2D Object Retrieval Based on Points and Curves

Like the previous chapter, this part of the thesis is also dedicated to the task of 2D object retrieval. Instead of using a pure skeleton-driven approach, this project is focusing on other shape descriptors like *contour points* and *contour curves* [Fei+14c; Yan+15a]. Both primitives are obtained by developing a sophisticated approach for the purpose of shape analysis, whereas the feature sets are either adapted (cf. [BMP02]) or reused (cf. [Yan+15b]). The actual matching is then realised by the Hungarian method which determines the best fitting correspondence configuration. It is worth knowing that the matching is performed twice due to the fact that both feature types are considered independently. Thus, two shape dissimilarity values are produced, one for the points and one for the curves. In order to retrieve the overall shape distance, both values are merged by a weighted sum.

By creating a setup highly similar to that which is known from the PSSGM approach, the performance of the skeletal structure can be evaluated towards other shape representations. Such an opposing comparison is valuable as a basis of argumentation as well as for the derivation of further actions in context of this work. However, considering the fact that the work of [BL08] has already proposed excellent results, the expectation is not to obtain a better score. The actual goal is to investigate how much effort is required to retrieve a similar result based on other shape descriptors, e.g. points and curves. In addition to this, the method is thoroughly tested by a qualitative and quantitative evaluation.

4.1   Fundamental Concepts

4.1.1   Douglas-Peucker Algorithm

The Douglas-Peucker algorithm [DP73] is a well-known method in the field of image processing and pattern recognition which is typically used to reduce the number of points forming an arbitrary curve. The technique itself is straightforward and easy to understand. In context of this project, the method is used to decrease the noise-to-signal ratio of the objects' contours used here for evaluation. Figure 4.1 demonstrates the working principle of the Douglas-Peucker algorithm by performing three iterations on an arbitrary line segment. In more detail, the algorithm requires a user-defined tolerance value ϵ which determines the

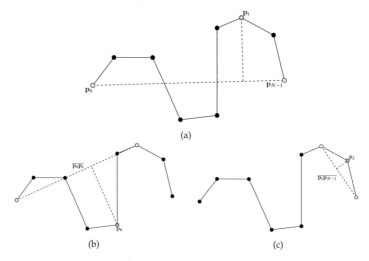

(a)

(b)   (c)

Figure 4.1: The figure shows the working principle of the Douglas-Peucker algorithm by illustrating three iterations on an arbitrary line segment. During the first iteration (Figure 4.1a) the entire line segment is used. The two dotted lines indicate the connection between the start and end point (white circles) and the maximum distance of this line to the point $\mathbf{p}_j$. Since its distance is above the tolerance value, the point is kept and drawn in grey. Afterwards, the line segment is split as demonstrated in Figure 4.1b and Figure 4.1c. Both sub-segments are then used to re-invoke the algorithm. Please notice that the point $\mathbf{p}_j$ in Figure 4.1c is below the given threshold ϵ and thus the point can be removed.

maximum degree of the simplification accepted by the user. Given ϵ as well as an *ordered* set of points $\{\mathbf{p}_0, \mathbf{p}_1, \ldots, \mathbf{p}_{N-1}\}$ representing the original curve, the method starts to remove those points which are not significantly contributing to its appearance. Therefore, the point set is recursively divided into subsets beginning with the start $\mathbf{p}_0$ and end point $\mathbf{p}_{N-1}$. Both points are then connected by a line and the point $\mathbf{p}_i$ (with $\mathbf{p}_i \neq \mathbf{p}_0 \neq \mathbf{p}_{N-1}$) is searched, which has the highest distance to $\overline{\mathbf{p}_0 \mathbf{p}_{N-1}}$. If this distance is below the tolerance value ϵ, the point is removed from the set. The algorithm processes recursively by creating two new line segments $\{\overline{\mathbf{p}_0 \mathbf{p}_i}, \overline{\mathbf{p}_i \mathbf{p}_{N-1}}\}$ which are used as new input.

4.1.2 Fourier Transform - A Brief Introduction

The Fourier Transform (FT)[1] or rather the Fast Fourier Transform (FFT) is a popular instrument originally coming from the field of signal processing where it is widely used to map a time-sampled signal to its frequency spectrum and vice versa. The FT was proposed by Joseph Fourier who recognised that every periodic function can be expressed as the sum of sine and cosine functions. Both representations are equivalent to each other and are convertible without any loss of information. In the context of image processing, the FFT operates on the spatial domain in order to transform the image's intensity values to its frequency spectrum. Using the spectrum of an image allows either to easily suppress or emphasise certain frequencies. Strictly speaking, low frequencies are representing smooth areas inside the image (or signal) whereas details in form of intensity edges are carried by high frequencies.

In the following, the equations are given to transform a (discretised) 1D time-signal to its Fourier spectrum. Please bear in mind that 2D signals are processed identically only by performing the 1D equations twice. For further details, the reader is referred to the book [Sch05] which also provides the equation depicted below. The actual transformation is then realised by taking the following basis vectors $\mathbf{e}_j$ which are corresponding to the frequency $j = 0, 1, \ldots, N-1$ (with N being the signal length):

$$
\begin{aligned}
\mathbf{e}_j(x) &= \frac{1}{\sqrt{N}} \exp\left(\frac{i2\pi j x}{N}\right) \\
&= \frac{1}{\sqrt{N}} \cos \frac{2\pi j x}{N} + \frac{1}{\sqrt{N}} i \sin \frac{2\pi j x}{N} \quad ,
\end{aligned}
\tag{4.1}
$$

where $x = 0, \ldots, N-1$. The interpretation is as follows: Due to the term $\frac{2\pi}{N}$ inside the sine and the cosine function, each $\mathbf{e}_j$ determines the number of full oscillations with respect to j. The actual transformation of the discretised function $\mathbf{f}(x)$ is then accomplished by applying each $\mathbf{e}_j(x)$ to $\mathbf{f}(x)$ in form a dot product (cf. [Sch05]):

---

[1]http://mathworld.wolfram.com/FourierTransform.html, [online: 19th August 2015]

$$F(j) = (re_j, im_j) = \langle \mathbf{f}^N(x), \mathbf{e}_j(x) \rangle \quad . \tag{4.2}$$

Please observe that the notation of the function $f(\cdot)$ changed to that of a vector $\mathbf{f}$ in order to stay consistent in presence of the dot product. Nevertheless, the equation above delivers the desired frequency spectrum as a sequence of complex numbers, describing the relation of cosine and sine that contributes to a certain frequency. For the sake of completeness, Equation (4.3) and Equation (4.4) illustrate the complete FT and its inverse:

$$F(j) = \frac{1}{\sqrt{N}} \sum_{x=0}^{N-1} f(x) \exp\left(-\frac{i2\pi jx}{N}\right) \quad , \tag{4.3}$$

where $f(x)$ returns the element of the (discretised) time-sampled function $f(\cdot)$ at position x which is more intuitive as vector notation: $\mathbf{f}(x)$ or rather $\mathbf{f}_x$. However, the more typical case is the continuous one and thus, the former expression is kept in the following.

$$f(x) = \frac{1}{\sqrt{N}} \sum_{j}^{N-1} F(j) \exp\left(\frac{i2\pi jx}{N}\right) \quad . \tag{4.4}$$

In the context of this project, the frequency spectrum is used to achieve the following goals: First, by applying an Low Pass Filter (LPF) to the spectrum the signal noise shall be suppressed. Second, the LPF is designed to emphasise those signal characteristics which are distinctive for the instances of the same object class. Figure 4.2 demonstrates the effect of noise suppression based on a randomly generated signal.

The FT has plenty of great advantages regarding signal processing tasks. Besides its native property of being horizontally invariant to translation, it additionally constitutes a powerful instrument in the area of data compression. Moreover, it is worth mentioning that the frequency spectrum is symmetric in cases where the signal samples are real. In more detail, the spectrum can be reduced by the half of its coefficients without any loss of information (cf. [Sch05]).

4.2   Contour Partitioning

This section describes the contour partitioning approach employed in this project. Therefore, the method first identifies a meaningful set of boundary points determining the most characteristic shape regions. Taking this set of points, the boundary of an object Ω can be separated into multiple contour segments. As described later in this chapter, this set of shape

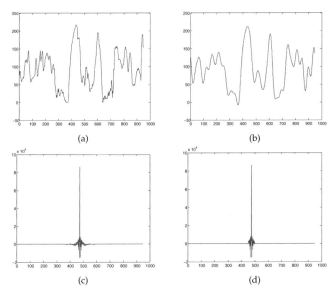

(a)   (b)

(c)   (d)

Figure 4.2: The figure illustrates the application of an LPF on an arbitrary signal (Figure 4.2a). The actual smoothing is implemented based on its frequency spectrum (Figure 4.2c). The shown spectrum is shifted in such a way that its low frequencies are centred. The chosen filter is designed to suppress the high frequencies around this centre by setting them to zero inside the frequency domain (Figure 4.2d). Finally, the smoothed signal is shown in (Figure 4.2b). **(Upper row:)** x-axis - time, y-axis - signal value; **(Lower row:)** x-axis - frequency band, y-axis - amplitude.

fragments is going to be exploited during the feature generation and thus, for the purpose of matching. The general idea is to detect contour regions having a high curvature toward the overall shape trend. Therefore, reference points q_i ($i > 0$) are placed inside the shape to compute the distance of all contour points p_j to their closest reference q_i. By traversing the shape sequentially in one direction, a signal is generated that can be used to detect peaks in its second-order derivative. Using these peaks, the desired partitioning points are obtained by a simple backtracking procedure. In the following, the single stages of this processing line are the subject of a detailed discussion. First, the object's contour has to be extracted in order to generate a polygon $\mathcal{P}$ of it. For the purpose of reducing noise artefacts, the Douglas-Peucker technique (cf. Section 4.1.1) is recursively applied to $\mathcal{P}$. Here only a small tolerance value $\epsilon^{(DP)}$ is taken which drastically reduces the noise-to-signal ratio without suppressing

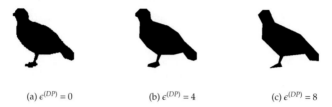

(a) $\epsilon^{(DP)} = 0$ (b) $\epsilon^{(DP)} = 4$ (c) $\epsilon^{(DP)} = 8$

Figure 4.3: The figure shows two smoothing results using the Douglas-Peucker algorithm. **(Left:)** Original shape polygon as expected result after being extracted from the image. **(Centre:)** Smoothing result by employing the Douglas-Peucker method to the original shape. **(Right:)** Smoothing result with a threshold level that is twice as high compared to the previous one.

significant properties of the shape. Figure 4.3 illustrates the smoothing on an exemplary image of a bird and two different tolerance values $\epsilon^{(DP)} = \{4, 8\}$. Subsequently, the previously mentioned reference points (q_i) are determined in order to generate a point-to-boundary distance signal. These reference points are then located by employing an FMM as proposed in [HF07a]. According to the description given in Section 2.2.6, the method generates a distance or rather a time-crossing map T in relation to a given initialisation. By exploiting the $\mathcal{P}$ as seed configuration, a process similar to a contour contraction can be realised. The result of this evolutionary procedure is illustrated in Figure 4.4. Finally, the searched q_i are

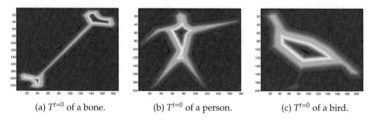

(a) $T^{t=0}$ of a bone.   (b) $T^{t=0}$ of a person.   (c) $T^{t=0}$ of a bird.

Figure 4.4: The figure shows three exemplary results of the FMM based on the polygonal boundary shape given as input to the algorithm. The actual distance information is depicted by the colour of the pixels (*dark blue* is the minimum and *dark red* the maximum). More interesting is the fact that the reference points are located inside the red coloured islands.

discovered iteratively at those value locations in T with maximum distance to the boundary. This iterative working principle is based on a dynamically adapting threshold $\mu^{(BG)}$ which is applied to T in order to remove all background values below a certain border with the intent to emphasise possible candidate regions:

$$\mu^{(BG)} = g(T(\Omega)) - 2 \cdot f(T(\Omega)) \quad , \tag{4.5}$$

where $g(\cdot)$ returns the maximum value: $\{x^\star \mid x, x^\star \in T, \, x, x^\star \in \Omega, \, \forall x \in T : x^\star \geq x\}$. In contrast to this, $f(\cdot)$ indicates the standard deviation. Once the background pixels are removed, the remaining values are considered as disjoint regions $\mathbf{A}_i$. For each $\mathbf{A}_i$ a *weighted centroid* (a_i) is determined based on the corresponding entries in T and their pixel locations. While collecting all a_i inside the set $\mathcal{K}^{t=0}$, the iteration is completed by creating a binary mask $\mathbf{B}$ which has the same size as the input image. This binary mask is then used to store all $\mathbf{A}_i^{t=0}$ (cf. Figure 4.5d).

Depending on the number of reference points found in the last iteration, $\|\mathcal{K}^{t-1}\|$ further cycles have to be performed. Therefore, each iteration is taking one of the points from $\mathcal{K}^{t-1}$ to initialise the FMM (cf. Figure 4.5b). In consequence of this, each iteration produces a distance map $\overrightarrow{T}_i^t$ (with $i = [1, \dots, \|\mathcal{K}^{t-1}\|]$). The notion is now to stress those object areas having the highest distance to the currently selected reference point by multiplying $\overrightarrow{T}_i^t$ with $T^{t=0}$. The result is illustrated in Figure 4.5 based on the example of a glass. As observable in Figure 4.5a and Figure 4.5d, the lower reference point is only detected by stressing the lower part of the object $T^{t=0}$ (cf. Figure 4.5c). Having the new distance map $T^t = \overrightarrow{T}^t \circ T^{t=0}$ (where $\circ$ indicates a pixelwise multiplication), the threshold $\mu^{(BG)}$ is adapted to localise further regions of interest ($\mathbf{A}_i^t$). Subsequently, the weighted centroids ($\mathbf{q}_i^t$) are calculated for each $\mathbf{A}_i^t$. In order to prevent duplicates residing close to each other, each further $\mathbf{q}_i^t$ has to pass a validation process. By taking the binary mask from the last iteration, the reference point $\mathbf{q}_i^t$ can easily be rejected if it is part of an already processed area: $\mathbf{B}^{t-1}(\mathbf{q}_i^t) = 1$. Otherwise, $\mathbf{q}_i^t$ is added to $\mathcal{K}^t$ and $\mathbf{B}^{t-1}$ is updated. The algorithm terminates if the current iteration returns an empty set of new references ($\mathcal{K} = \emptyset$). The final partition set is then generated by: $\bigcup_{j=0}^t \mathcal{K}^{t=j}$. Having this set, a last filtering run is required to delete false positives which were not recognised during their detection. Therefore, each reference point has to pass two further conditions: First, all points have to satisfy a certain distance to the shape's centre of gravity. If only one point is violating this condition, all reference points are discarded expect the one responsible. Second, all references are removed whose boundary parts are split into multiple segments. Strictly speaking, the assigned boundary segments have to be fully connected. Once the validation is completed, the reference-to-boundary distance signal can be generated by creating a further time-crossing map based on the final set of references. Taking this map, only those values are selected which are covered by the object's

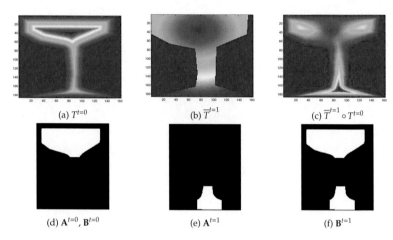

(a) $T^{t=0}$ (b) $\overline{T}^{t=1}$ (c) $\overline{T}^{t=1} \circ T^{t=0}$

(d) $\mathbf{A}^{t=0}$, $\mathbf{B}^{t=0}$ (e) $\mathbf{A}^{t=1}$ (f) $\mathbf{B}^{t=1}$

Figure 4.5: The figure demonstrates the working principle of the proposed method based on several important stages. Figure 4.5a and 4.5c show the distance maps which are exploited for localising the reference points inside the object. Figure 4.5b illustrates the time-crossing map used for stressing the lower object part. In contrast to this, Figure 4.5d and 4.5e display the regions which have been preserved after the background deletion in 4.5a and 4.5c. While 4.5d constitutes the binary mask of the first cycle, its counterpart in the next iteration (4.5f) is obtained by combining 4.5d and 4.5e.

contour $\partial\Omega$ in order to arrange them inside a vector $\mathbf{e}$. The number of elements in this vector equals the amount of pixels forming the shape's contour. By considering $\mathbf{e}$ as 1D signal, the goal is to derive its second-order derivative to determine the maximum and minimum points. By backtracking these peaks to the contour pixels, breaking points are found to separate the object's contour into sub-parts. Figure 4.6 shows both the contour-masked time-crossing map and its corresponding reference-to-boundary distance signal. Since the bone structure yields two reference points, the contour is separated into two parts which are shown in the first and the second column. Moreover, it illustrates a number of regularities impeding this peak detection. A problem which is primarily caused by noise with the result of an imprecise contour partition set. Such a situation also affects all subsequent matching approaches. In order to alleviate artefacts and to increase the robustness of the proposed method, the signal is further smoothed by the FFT (cf. Section 4.1.2). By transforming $\mathbf{e}$ to its frequency spectrum $\hat{\mathbf{e}}$, a Low Pass Filter (LPF) can be implemented by suppressing the high frequencies which are carrying the noise. This way the noise-to-signal ratio is decreased without losing significant characteristic of the signal or rather the shape. High frequencies

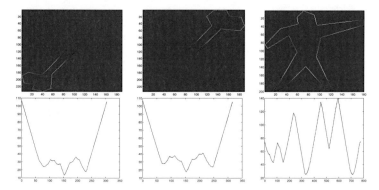

Figure 4.6: The figure shows two example objects, a bone and a person, whose boundary information carries the distance to the closest reference point (upper row). Please notice that the bone structure yields two reference points and thus, the contour is split into two fragments (first and second column). The blue colour indicates a low distance of the shape segments to the references assigned to them. The bottom row illustrates the corresponding non-smoothed reference-to-boundary distance signal, where the x-axis depicts the pixel location and the y-axis its distance to the reference point.

are typically transporting the very fine details of a shape and thus they are not relevant and even undesired for the actual goal to separate the object's boundary into segments. This behaviour is demonstrated in Figure 4.7, where the signals of Figure 4.6 have been smoothed by applying an LPF to them. Here the LPF is established as a mask of Gaussian coefficients centred at the zero-shifted frequency domain. By invoking such a filter design on the spectrum, the amplitudes values are not completely erased, instead they are weighted and prioritising the low frequencies. Two things have to be noted here: First, in order to alleviate border artefacts, the spectrum has to be padded at its start and its end before it can be passed to the LPF. Second, depending on the number of reference points, the padding process is behaving differently. In other words, in presence of a single reference point, the spectrum is only doubled, whereas it has to be additionally flipped if more than only one reference has been detected. This can easily be explained due to the fact that the contour is not separated into different subsets if only one point exists. Assuming that the LPF could be applied successfully to the spectrum $\hat{e}$, the first ($\check{e}'$) and the second-order ($\check{e}''$) derivatives are calculated based on its smoothed version $\check{e}$. Afterwards, this data is exploited to determine all zero locations in $\check{e}'$ having either a positive or negative curvature. The latter information can be gathered from $\check{e}''$. Figure 4.8 shows the second-order derivatives using the signals

Figure 4.7: The figure corresponds to Figure 4.6 and presents the same reference-to-boundary distance as before. However, the signal of the bone and the person has been smoothed by applying an LPF to them. In order to alleviate border artefacts, the signal has been mirrored to pad its start and its end. The original data is visually separated by the dotted lines.

from Figure 4.7. Please bear in mind that the illustrated second-order derivatives are only considering a signed binary version of the first-order ones: sign($\breve{e}'$), with sign($\cdot$) returning 0.0 if $\breve{e}'_x = 0.0$, 1.0 if $\breve{e}'_x > 0.0$ and otherwise −1.0). Having this information, the desired partition points can easily be determined by back-tracking the peaks (those locations in $\breve{e}''$ unequal zero) to the original signal and then back to the shape's contour.

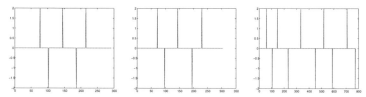

Figure 4.8: The figure illustrates the second-order derivatives which have been obtained by taking the smoothed signals shown in Figure 4.7. Please notice that the plots of these derivatives have been generated by only considering a signed binary version of the first-order ones. Having this information, the desired partition points can easily be determined by back-tracking the peaks to the original signal and then to the shape's contour.

In practise, this approach runs the risk of still encompassing a limited amount of false positives due to an inappropriate set of Gaussian smoothing coefficients. Unfortunately, there is no general solution for these coefficients which fits all object instances due to varying degrees of deformation and noise. In order to tackle this problem, the second-order derivative has to be analysed in terms of closely located neighbours having a small distance to each other. Figure 4.9 demonstrates such a situation, where the red circle indicates the problem. If such a situation is detected, the peaks height is taken from $\breve{e}$ of the two affected positions in order to compute its difference. If this deviation is also small, the Gaussian coefficients are adapted by reducing its standard deviation. Afterwards, the smoothing

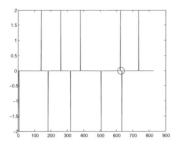 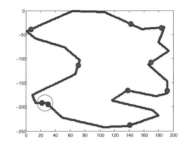

Figure 4.9: The figure illustrates a situation in which a false positive was generated due to an inappropriate set of Gaussian smoothing coefficients. These artefacts can easily be determined by analysing the second-order derivative. The red circles indicate both the region inside the derivative responsible for the falsely detected partition point and its consequence towards the aim to separate the contour into meaningful segments.

procedure is repeated based on the original **e**. Experiments showed that typically only one to two iterations are required to obtain a reliable set of partition points. Figure 4.10 shows the final results of the bone, the person and a bird.

 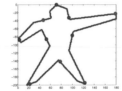 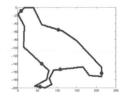

Figure 4.10: The figure illustrates the finally obtained contour partition set of the bone, the person and a bird. The contour is then separated into multiple segments which are respectively defined by the pairs of partition.

4.3 Contour Curve Segment-driven Feature Generation

This part introduces the descriptors which are employed for the similarity calculation and thus, for the implementation of the object retrieval process. Specifically, the feature set is divided into two components: One part is taking into account the partition points introduced in Section 4.2, whereas the second one is only operating on the contour segments which have been obtained by exploiting the partition points. Both sets are considered independently from each other and are finally merged during the matching in form of a weighted sum.

While the former set can be viewed as an adapted derivative of the well known Shape Context descriptors proposed in [BMP02], the latter feature type has first been presented by Cong et al. in [Yan+14]. Both of them are going to be discussed in more detail below.

4.3.1   The Point Context Shape Descriptor

The name *Point Context (PC)* as well as the actual idea behind this descriptor is derived from the so-called *Shape Context (SC)* which has been introduced by Belongie et al. in [BMP02]. The Point Context, in turn, has been published by Feinen et al. in [Fei+14c]. The term Point Context originates from the fact that it is directly applied to the partition points presented in Section 4.2. Although both approaches are quite similar to each other, the Point Context descriptor goes without any alignment transformation (cf. [BMP02]).

Shape Context   A short excursion into the subject of the so-called SC is inevitable since it provides the foundation that was adapted for the development of the PC. Recalling that the PC is co-responsible for the final establishment of correspondences, a comprehensive knowledge about its roots supports a better interpretation of the corresponding evaluation results.

> "[The Shape] descriptor, the Shape Context, [is used] to describe the coarse distribution of the rest of the shape with respect to a given point on the shape. Finding correspondences between two shapes is then equivalent to finding for each sample point on one shape the sample point on the other that has the most similar Shape Context" [BMP02]

Although the SC has an intuitive and easy to understand working principle, it is a powerful instrument for the purpose of shape recognition and categorisation as demonstrated in [BMP02]. Additionally, this work confirms the property of being highly distinguishable in form of its PC descriptor. Moreover, the SC has the capability of involving other sources to increase its discrimination power. With the intention of generating the SC, the object's boundary $\partial\Omega$ is considered as a discrete set of points. Therefore, the boundary is sampled with a certain number of uniformly distributed points $v_i \in \mathbb{R}^2$. These sample locations are not forced to represent significant shape characteristics as e.g. the maxima of curvature. If the number of sample points is sufficiently large and the contours are piecewise smooth, a good approximation of the shape can be expected. The SC is implemented as a histogram $\mathbf{H}$ which is carrying the relative positions of the sample points. By selecting a certain v_i from the entire set of samples $\mathbb{D}$, the histogram $\mathbf{H}_i$ encodes the relative distribution of the remaining sample points $\mathbb{D}/\{v_i\}$ in respect to v_i. In order to get a higher sensitivity close to

v_i, the histogram is defined inside the log-polar space. This increased sensitivity provides an elegant and efficient way to alleviate the linearly growing positional uncertainty towards an increasing distance to v_i (cf. [BMP02]).

Point Context In comparison to the SC, the Point Context mostly differs in terms of its specialised input data and its missing alignment transformation. In order to obtain deeper insights, the reader is referred to [BMP02]. According to the implementation of the SC, the boundary has to be sampled by a fixed number of sample points $v_i \in \mathbb{R}^2$. There is no upper limit and thus, all contour points are allowed to be taken into account. In contrast to the SC, the new descriptor is only generated for each partition point $\hat{p}_i$ that has been determined in Section 4.2. The relative position to all sample points v_i is then easily established by creating a set of direction vectors $\{v - \hat{p}_i\}$ which are originating in the reference point $(\hat{p}_i)$. As in the case of the SC, these relative positions are mapped into the log-polar space in order to be more sensitive close to the currently selected reference point. Afterwards, a histogram H_i is calculated carrying the configuration of the entire shape relative to p_i (cf. [BMP02]):

$$H_i^{(r,\theta)} = \#(\{v \neq \hat{p}_i : (v - \hat{p}_i) \in bin_{lp}(r,\theta)\}) \quad , \tag{4.6}$$

where $\#(\cdot)$ is a function that counts the elements in a set, while r and θ indicate the histogram bins with relation to the log-polar space. In order to be scale invariant, all radial distances are normalised based on the average distance between all sample point pairs. The actual matching is then performed by finding the best fitting correspondence configuration between the partition points of both shapes. Therefore, the 2D histogram is transformed into a sequential aligned vector by concatenating the rows of $H_i^{(r,\theta)}$. Strictly speaking, if the histogram encompasses N bins for r and M bins for θ, the resulting feature vector will be:

$$\hat{x}_i^{(PC)} = (x_1, x_2, \ldots, x_j, \ldots, x_{NM})^\top \quad , \tag{4.7}$$

with $x_j \in H_i^{(r,\theta)}$.

4.3.2 The Contour Segment Shape Descriptor

The contour segment representation has been introduced by Cong et al. in [Yan+14]. In context of this work, this shape representation is recruited to be fused with the Point Context descriptor for the purpose of implementing a robust object retrieval system. By taking the set of partition points, the object contour can be divided into multiple contour segments which are then respectively described by exploiting a 10-dimensional feature vector $\hat{x}^{(CS)}$:

$$\hat{\mathbf{x}}^{(CS)} = (x_1, x_2, \ldots, x_j, \ldots, x_{10})^\top \quad . \tag{4.8}$$

Each vector element describes the currently observed contour fragment by a geometrical relation which can be derived from it. In the following, these quantities are introduced in more detail. The features are quite simple and can be reconstructed without great effort. Starting with the first element in $\hat{\mathbf{x}}^{(CS)}$, the discussion will successively cover all subsequent elements. Thus, the value of $x_1^{(CS)}$ simply constitutes the Euclidean distance of a virtual line spanned between the start and end point of a segment. The total number of pixels along the contour segment (CS) is encoded in $x_2^{(CS)}$. Merging these two features in $x_3^{(CS)}$, an indicator is obtained expressing the degree of deviation between the shape of the CS and a straight line. This ratio is primarily dedicated to the detection of path variations. In order to distinguish contour fragments with the structure demonstrated in Figure 4.11a from those in 4.11b, the enclosing area between the virtual line and the actual contour (grey areas in Figure 4.11) is determined as feature value ($x_4^{(CS)}$).

(a)   (b)

Figure 4.11: The figure illustrates two contour fragments which are equal concerning their feature values $x_1^{(CS)}$ and $x_2^{(CS)}$. In order to distinguish between these two segments, $x_4^{(CS)}$ has been introduced to determine the enclosing area depicted in grey.

While the first four elements are quite intuitive, the remaining ones are more sophisticated ratio relations. For the purpose of being rotation invariant, the contour segments are first normalised in their pose, before the remaining feature values are calculated. Therefore, each CS is rotated vertically until start and end point are residing on the y-axis ($\hat{\mathbf{y}}$) of the world coordinate system. Although the underlying transformation is quite simple, the result is ambiguous towards its alignment on the x-axis ($\hat{\mathbf{x}}$). Strictly speaking, the fragment can be mirrored at $\hat{\mathbf{y}}$. With the intent to tackle this problem and to resolve this ambiguity, the method ensures that the majority of contour points are oriented to the right hand side of $\hat{\mathbf{y}}$. In consequence of this, all contour elements are aligned rotation invariant for the actual feature generation (cf. Figure 4.12). Taking these pose corrected CS, the remaining feature values are calculated based on a bounding box-guided scheme:

$$
\begin{aligned}
x_5^{(CS)} &= h/w \\
x_6^{(CS)} &= h'/w_1' \\
x_7^{(CS)} &= h'/w_2' \\
x_8^{(CS)} &= h'/w_3' \\
x_9^{(CS)} &= (w_3'h')/(w_1'h') \\
x_{10}^{(CS)} &= (w_2'h')/(w_1'h')
\end{aligned}
\qquad , \qquad (4.9)
$$

where h and w are the corresponding height and width of the global bounding box which entirely surrounds the CS. Moreover, this bounding box is separated into three subareas of equal height which are depicted by the symbols h' and $w_{\{1,2,3\}}'$. This arrangement is also addressed in Figure 4.12 in order to clarify this correlation. The feature generation process terminates with a further normalisation step of the feature vector $\hat{\mathbf{x}}^{(CS)}$ with the intention of becoming scale invariant. This is realised by employing the half of the bounding box perimeter as denominator for each vector element $x_i^{(CS)}$:

$$
\hat{\mathbf{x}}^{(CS)} = \frac{\hat{\mathbf{x}}^{(CS)}}{w+h} = \left(\frac{x_1}{w+h}, \frac{x_2}{w+h}, \ldots, \frac{x_{10}}{w+h} \right)^{\top} \quad . \qquad (4.10)
$$

4.4 Object Matching Based on Points and Curves

After introducing two complementary feature types, this section addresses the actual matching procedure. While both feature types are generated independently from each other, they are combined for the purpose of finding the best fitting correspondence configuration. Being aware of the fact that both feature sets count to the class of *shape descriptors*, the proposed recognition approach aims at the *categorisation* of 2D objects.

4.4.1 Graph Matching by the Hungarian Method

The bipartite graph is a classical instrument to realise matching tasks by passing it e.g. to the Hungarian method which is capable of determining node correspondence with minimum costs. Both the bipartite graph and the Hungarian method are subject of Chapter 2.

As a quick reminder, the goal of the matching is to determine alignments between the nodes of two graphs G and G'. Therefore, each node pair is characterised by a cost value which is calculated based on a predefined dissimilarity measure. This concept is also employed in this project for the purpose of finding best fitting correspondence configurations

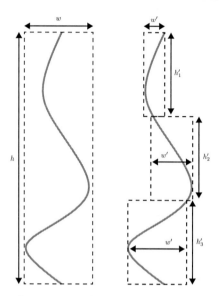

Figure 4.12: The figure shows an exemplary contour segment which has been normalised in its pose by aligning the start and end point vertically so that both points are collinear towards the y-axis of the world coordinate system. Moreover, it illustrates the separation of the global bounding box into three subareas. **(Left:)** Global bounding box which surrounds the entire segment. **(Right:)** Subareas which have been derived from the global one by dividing it into three parts having the same height.

to derive the overall dissimilarity between two objects. Having in mind that this project utilises two types of features, namely the Point Context (PC) and a contour segment (CS) descriptor, the dissimilarity measure used must be capable of combining the outcome of both. Strictly speaking, the overall matching costs are calculated as weighted sum of the PC and the CS descriptor. In the following content, it is assumed that $\partial\Omega$ and $\partial\Omega'$ are two different contours which have been extracted from the objects Ω and Ω' to be matched.

4.4.1.1   Shape Dissimilarity Based on the Point Context Descriptor

As already explained in Section 4.3.1, the PC descriptors are generated based on a set of partition points which have been extracted during the preprocessing. These points are residing on the objects' boundaries: $\mathbf{p}_i \in \partial\Omega$ ($i = [1,2,\dots,N]$) and $\mathbf{p}'_j \in \partial\Omega'$ ($j = [1,2,\dots,K]$), respectively, with $N \leq K$. In analogy to the Shape Context (SC) [BMP02], the dissimilarity or rather costs ($c(\cdot,\cdot)$) between two partition points $\mathbf{p}_i$ and $\mathbf{p}'_j$ is calculated based on their PC histograms $\mathbf{H}_i$ and $\mathbf{H}'_j$ coupled with χ^2 test statistic:

$$c(\mathbf{p}_i, \mathbf{p}'_j) = \frac{1}{2} \sum_{k=1}^{M} \frac{[\hat{\mathbf{x}}^{(PC)}_{i,k} - \hat{\mathbf{x}}'^{(PC)}_{j,k}]^2}{\hat{\mathbf{x}}^{(PC)}_{i,k} + \hat{\mathbf{x}}'^{(PC)}_{j,k}} \quad , \tag{4.11}$$

where M depicts the length of the feature vector or rather the sequentially aligned histogram rows. Taking this dissimilarity function, the matching costs of all partition point pairs can be arranged in an overall cost matrix $\mathbf{C}$:

$$\mathbf{C} = \begin{pmatrix} c(\mathbf{p}_1, \mathbf{p}'_1) & c(\mathbf{p}_1, \mathbf{p}'_2) & \cdots & c(\mathbf{p}_1, \mathbf{p}'_K) \\ c(\mathbf{p}_2, \mathbf{p}'_1) & c(\mathbf{p}_2, \mathbf{p}'_2) & \cdots & c(\mathbf{p}_2, \mathbf{p}'_K) \\ \vdots & \vdots & \ddots & \vdots \\ c(\mathbf{p}_N, \mathbf{p}'_1) & c(\mathbf{p}_N, \mathbf{p}'_2) & \cdots & c(\mathbf{p}_N, \mathbf{p}'_K) \end{pmatrix} \quad . \tag{4.12}$$

Finally, this cost matrix can be passed to the Hungarian method determining the best fitting alignment between both sets of partition points. A typical problem which frequently occurs in the context of using the Hungarian method is the necessity to incorporate further dummy nodes to the matching process. This requirement emerges whenever both partition point sets do not encompass the same number of elements which is most probably the case for many object pairs. Only if both sets are balanced in their size, a one-to-one matching can be performed. This problem can be solved by incorporating dummy nodes which assign a penalty to all nodes which are not able to find an appropriate partner during the matching. The actual implementation of such a solution is just the insertion of multiple additional rows of either constants [BMP02] or dynamic values [BL08] to square the shape of $\mathbf{C}$.

Being aware of this, the overall shape dissimilarity can be calculated as the average or sum of all matching costs (or penalties) being part of the final correspondence configuration. Let $c_1, c_2, \ldots, c_K$ be the short notation for the cost values of the final alignment, then the overall shape dissimilarity can be calculated as follows:

$$c^{(PC)} = \frac{1}{K} \sum_{i=1}^{K} c_i \quad . \tag{4.13}$$

4.4.1.2 Shape Dissimilarity Based on the Contour Segment Descriptor

Similar to the PC, the matching is realised in context of the CS descriptor. Therefore, let C_i and C'_j be two contour segments which are part of $\partial\Omega$ and $\partial\Omega'$, respectively. The numbers of CSs equals the amount of partition points which have been detected during the preprocessing of both objects. Thus, the setup is identical to the one before and there exist as many curve fragments as partition points: $i = [1,2,...,N]$ and $j = [1,2,...,K]$, respectively, with $N \leq K$. Equivalent to the PC, a dissimilarity measure is required to retrieve the matching costs between two CSs:

$$c(C_i, C'_j) = \frac{1}{M} \sum_{m=1}^{M} \frac{\sigma_n |\hat{\mathbf{x}}_{i,m}^{(CS)} - \hat{\mathbf{x}}_{j,m}'^{(CS)}|}{\sum_{k=1}^{K} |\hat{\mathbf{x}}_{i,m}^{(CS)} - \hat{\mathbf{x}}_{k,m}'^{(CS)}|} \quad , \tag{4.14}$$

where $\hat{\mathbf{x}}_i^{(CS)} \in \mathbb{R}^M$ and $\hat{\mathbf{x}}_j'^{(CS)} \in \mathbb{R}^M$ are the features vectors of C_i and C'_j introduced in Section 4.3.2. Please notice that $c(C_i, C'_j)$ involves the entire shape in order to calculate the distance between two contour segments. Furthermore, the parameter σ_m has to be interpreted as weight for each feature. It has been derived by a sophisticated optimisation process which is thoroughly explained in [Yan+15b]. Beyond that, this distance measure requires the clockwise traversal of the feature vectors according to how their corresponding contour segments occur by traversing the objects boundary. Thus, Equation (4.14) is not symmetric in terms of $c(C_i, C'_j) \neq c(C'_j, C_i)$.

Knowing the distance measure, the matching procedure continues to create the overall cost matrix **C** (like in context of the PC), in order to fulfil the goal of finding that alignment leading to minimum matching costs:

$$\mathbf{C} = \begin{pmatrix} c(C_1, C'_1) & c(C_1, C'_2) & \cdots & c(C_1, C'_K) \\ c(C_2, C'_1) & c(C_2, C'_2) & \cdots & c(C_2, C'_K) \\ \vdots & \vdots & \ddots & \vdots \\ c(C_N, C'_1) & c(C_N, C'_2) & \cdots & c(C_N, C'_K) \end{pmatrix} \quad . \tag{4.15}$$

The method completes according to the explanations already provided in Section 4.4.1.1. In cases where the number of CSs is different between the sets, dummy nodes have to be included to square the shape of the cost matrix **C**. Finally, the Hungarian method can be applied to the cost matrix given in Equation (4.15). Subsequently, the overall shape dissimilarity $c^{(CS)}$ is calculated by exploiting the resulting correspondence configuration. This computation is then performed in analogy to (4.13) where the average value is generated over the sum of all cost values $c_1, c_2, ..., c_K$ being part of the final alignment.

4.4.1.3 Overall Shape Dissimilarity Based on Points and Curves

Connecting to the discussion of calculating the shape dissimilarity either by the Point Context or the contour segment descriptor, this section shows how to combine both pieces of information. For this purpose, the following function is introduced:

$$c(\Omega,\Omega') = \alpha \, c^{(PC)} + (1-\alpha) \, c^{(CS)} \quad , \tag{4.16}$$

where the value of α is optimised by an integrated *Gradient Hill Climbing* [RN09] and a *Simulated Annealing* [KGV83] method. Please consult Section 8.2 to retrieve a more detailed insight concerning the concrete configuration of this parameter as well as information regarding the recognition performance of the proposed method.

4.5 Summary

After encountering skeletons in Chapter 3, this thesis part focuses on curves and contours for the purpose of 2D object recognition or rather categorisation. Therefore, the same data set is used for evaluation as already employed in context of the Path Similarity Skeleton Graph Matching (PSSGM). This way the discrimination power of the skeletal structures can be compared to other shape descriptors with the typical expectation that skeletons perform better, worse or almost similar. Here the situation is slightly different since the PSSGM has already delivered an almost perfect recognition performance. Thus, the assumption to accomplish a better score than it was low. However, even in the light of these expectations, the findings are extremely valuable for the entire work. Considering the situation that the proposed method obtains a score worse than the PSSGM, it would argue for the use of skeletons to realise object recognition. If both techniques performed almost equally to one another, it would not argue against the application of the skeletal structures.

In contrast to skeletons, this project exploits points and segments which are derived from the shape's contour. Therefore, a method is proposed that first determines a meaningful set of points residing on the boundary, the so-called *partition points*. Afterwards, the entire contour is divided into smaller segments at these locations. With access to these points and segments, descriptors can be generated to represent both for the purpose of object recognition. In order to guarantee almost identical sets of points and curves for instances of the same object class, a sophisticated technique has been developed. By additionally introducing so-called reference points inside the object (localised by means of a Fast Marching Method), a signal is generated encoding the distances from the reference(s) to the boundary. Having this signal, a Fourier Transform maps these values to a frequency spectrum, where the signal is smoothed and subsequently derived to its first- and second-order derivatives. Afterwards, maxima and minima are located inside these derivatives which finally yield the desired partition points.

Once this data is available, the goal is to find a feature set that represents this information appropriately. Therefore, two different approaches have been adapted or rather reused for the actual implementation. In context of the partition points, the well-known Shape Context [BMP02] has been modified with the outcome of a new descriptor, namely the *Point Context*. Since both descriptors are operating on the same main concept, their implementations are quite similar. Apart from this, the description of the contour segments is realised by geometry-driven features proposed by Cong et al. in [Yan+15b]. This feature collection encompasses ten quantities which characterise the shape of an arbitrary contour segment. The actual matching costs are then calculated based on these two sets. Subsequently, they are passed to the Hungarian method which, in turn, is responsible for finding the best fitting correspondence configuration. By calculating the average over the sum of all nodes which are part of the final matching, the shape distance between two objects is obtained. Please notice that the shape distance is calculated independently for the partition points and for the contour fragments, whereas the overall shape dissimilarity is gained as a weighted sum of both descriptors.

The recognition performance which has finally been accomplished during a comprehensive evaluation is only slightly worse than the one presented in context of the PSSGM. The approach is even better than the re-implementation of the PSSGM introduced in [Hed+13]. Moreover, it clearly outperforms other state-of-the-art techniques which are purely operating on contour segments [Yan+14]. Summarised, the proposed method is able to achieve excellent results by combining contour points and segments. However, the reliability of the partition points is of particular importance for the success of the proposed technique. Additionally, it is worth mentioning that the recognition performance of [Yan+14] has been increased drastically by merging the contour information with that of the skeletal structures. This synergy does even outperform the constellation of contour points and curves as presented above. Finally, it is interesting to compare the activities which were required to obtain these results. While the skeleton-based approaches only require the extraction of the skeletal structure as well as an adequate sampling scheme for its branches, the presented shape-driven technique takes more attention in order to behave robustly.

Chapter 5

3D Object Recognition Based on 3D Curves

While the last chapters have addressed shape matching in 2D, this section is devoted to an intensive study dealing with the topic of 3D object recognition. Therefore, a 3D object retrieval system is designed working in analogy to the Path Similarity Skeleton Graph Matching (PSSGM) as introduced in Chapter 3 (cf. [Fei+14b; Fei+14a]). For this purpose, modifications are necessary in order to apply the algorithm in 3D space. Hence, this project is dedicated to different aspects: Firstly, a representation like the skeletal structure in 2D is required, which quantifies significant visual object parts. Strongly connected to this, an efficient strategy is needed for sampling these perceptual regions meaningfully. Secondly, descriptors have to be introduced capable of encoding the geometrical characteristics of the underlying object appropriately. Thirdly, the project is concluded by a thorough evaluation which investigates the potential of this 3D version of the PSSGM.

Strictly speaking, the approach is operating on top of a depth acquisition system able to capture depth images from a scene. These images are subsequently passed to functional parts proposed in [Yi+13] and [MYL13]. Here it is worth mentioning that these depth maps are used to generate incomplete 3D point clouds for e.g. detecting the ground plane or deriving the 3D contour curves. Please notice that both the depth image and the point cloud are used to obtain information essential to perform robust object recognition.

The section starts with referring those works which are highly related and introduces basic concepts exploited in this project. Subsequently, the single stages of the processing pipeline are discussed in individual sections which encompass the following topics: (i) The estimation of a Local Coordinate System (LCS) responsible for navigation tasks inside the object (Section 5.3.1). (ii) The extraction of indicators (feature points) representing the most significant object parts (Section 5.3). (iii) The establishment of shortest paths as well as a meaningful sampling scheme. (iv) The introduction of path descriptors consisting of robust and distinctive quantities which additionally are rotation, scaling and translation (RST) invariant (Section 5.5). (v) The replacement of the Hungarian method with a new matching approach, called Maximum Weight Cliques (MWC) (Section 5.6).

5.1 Fundamental Concepts

5.1.1 Random Sample Consensus Method

The Random Sample Consensus (RANSAC) approach is a popular instrument in all areas where strong outliers badly influence a sequence of automatically measured data, e.g. in the area of active vision to optimise the estimation of fundamental matrix. RANSAC has been proposed by Bolles et al. in [BF81]. Roughly speaking, it iteratively estimates the best fitting parameter configuration of a model from a set of observed data similar to the Method of Least Squares (LSM). However, the Method of Least Squares (LSM) is not as robust as the RANSAC in presence of tremendous fraction of outliers.

In the following, the working principle of RANSAC shall briefly be described to enable a better interpretation of its result since it constitutes the foundation for the actual mapping task. For more detailed information, the work of [Zul12] is referred here as thorough introduction into this topic. However, for the following explanation, let $\mathbb{D}$ be an observation of unordered 2D data points and $\theta^\star$ those parameters which optimally fit a line into $\mathbb{D}$. Further assume that $\mathbb{D}$ encompasses enough valid data points to provide an optimal solution. In addition to this, the model space is over-determined and suffering from outliers. Given this situation, the algorithm starts to randomly pick up the minimum number of data points $\mathbb{D}^{c,t}$ required to establish the first hypothesis of the model with the parameter space θ^t. Here the dimension of θ equals two. Please notice that the algorithm does not incorporate the entire set of observations as it would be the case for the LSM. This property in particular reflects the power of RANSAC approach. Subsequently, the quality of θ^t has to be determined by taking into consideration those points of $\mathbb{D} \setminus \mathbb{D}^{c,t}$ which are below or equal to the standard deviation of correspondences, the *consensus set*. This procedure is repeated K times, where K is large to guarantee almost deterministic results in such a sense that all model configurations are close to each other if the method is repeatedly invoked on the same data set. Optionally, the data fitting method can finally be applied to the consensus set in order to retrieve the optimal solution $\theta^\star$. Figure 5.1 illustrates the expected results of both the RANSAC and LSM based on a toy example. Moreover, the figure shows the problem of the LSM which straightly operates on the entire measurement.

The result's accuracy of the algorithm highly depends on the selected values for standard

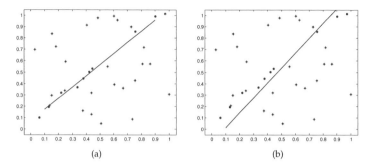

(a)   (b)

Figure 5.1: The figures show the same measurement which is suffering from a heavy set of outliers (blue stars). **(Left:)** The RANSAC is exploited to fit a line into this observation. **(Right:)** The LSM is employed. Both methods provide good estimations of the real data (red dots). Nevertheless, the RANSAC approach leads to a smaller error compared to the LSM, a fact which is also underlined by the visual impression.

deviation of inliers as well as for the number of iterations. The latter can be approximated with the following equation:

$$1 - p = (1 - a^b)^N \quad , \tag{5.1}$$

where p is the desired probability that at least one random set is chosen consisting of only inliers ($\mathbb{D}^\star$), this value is typically set to 0.99. Moreover, the symbol a depicts the probability of selecting inliers in general, whereas b and N respectively determine the number of model parameters and the required amount of iterations needed to retrieve at least one set $\mathbb{D}^\star$ with probability p.

5.2 Scene Analysis and 3D Curve Recovery

Two different methods are subject of the content below which are jointly used during the pre-processing in order to convert the raw data into meaningful structures. One method is published by the authors Yi et al. dedicated to the area of robot navigation. Its main contribution is a navigation framework enabling a "robot to continuously identify and [finding a] way toward a non-static target, e.g. following a walking person" [Yi+13]. In context of this project, the functionality is reused to automatically analyse a scene captured by a Kinect© device. The other method is presented by Ma et al. in [MYL13]. They introduce a procedure to describe 3D objects using 3D curve segments which approximate the object's underlying geometry. Comparing the method of this project to the latter one, both might seem to be quite similar to each other. However, they primarily differ towards the generation of the affinity graph and its corresponding weights.

5.2.1   Scene Analysis Concerning 3D Object Detection and Extraction

The detection and tracking of non-static objects is a challenging task requiring a sophisticated strategy. Therefore, the authors Yi et al. propose a method in [Yi+13] for simultaneous navigation and tracking. The latter in particular is of major interest for this work since it is using a so-called *Footprint Detection (FD) based Tracker* trained in an unsupervised process by analysing the scene's layout. The actual footprint detection is performed on a *plan-view map* as shown in Figure 5.2c. The FD-Tracker is also employed in this section, but for the purpose of 3D object recognition.

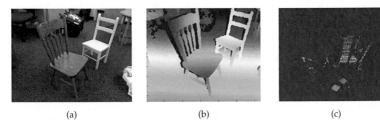

| (a) | (b) | (c) |

Figure 5.2: The figure gives a rough overview of the processing steps necessary to segment an arbitrary scene into 2D clusters which are representing the regions of interest. **(Left:)** Gray-scale image showing the scene, **(Centre:)** the corresponding depth image and **(Right:)** the scene captured from the top pointing upright to the ground plane.

"Then, the information is fused into a Bayesian tracking framework, generating very limited number of target candidates [in order to alleviate complexity issues]" [Yi+13]. In the following, the tracking principle shall briefly be described. By taking a Kinect$^{©}$ device, a depth image $I^{(depth)}$, as shown in Figure 5.2b, is acquired with the intent to estimate the ground plane of the scene. Therefore, it has to be enhanced in terms of a noise and holes reduction whose result is depicted with $\hat{I}^{(depth)}$. To further facilitate the detection of the floor plane, the largest surface inside the captured scene is assumed to be the desired one. It is not mandatory that the ground parts are entirely connected since the RANSAC approach is used to estimate the plane parameters: $\mathcal{K} = (\theta_r, \theta_p, \theta_h)^\top$ determining the camera's *tilt* (θ_r), *pitch* (θ_p) and *height* (θ_h). Therefore, $\hat{I}^{(depth)}$ is transformed into a 3D point cloud $\mathcal{P}^3$ and passed as input to the RANSAC approach. After a pre-defined number of iterations, the method returns the best fitting values for θ_r, θ_p and θ_h. Finally, all points of $\mathcal{P}^3$ are projected on $\mathcal{K}$, as illustrated in Figure 5.2c, enabling the procedure to segment multiple, disjoint areas. These 2D point clusters, generated based on the plan-view map, constitute the wanted footprints or rather 3D scene objects. For further details, the reader is referred to the original paper [Yi+13].

5.2.2 3D Object Representation by 3D Curves

After retrieving a set of object clusters extracted from the scene by the mechanism described above, these candidates have to be represented adequately for the purpose of matching. Therefore, the geometry of each scene object is approximated with multiple 3D curves. This description of 3D surfaces is taken from the work of the authors Ma et al. who employ this technique in their work [MYL13]. According to their observations, the use of curves is reasonable due to several advantages, namely the representation by a certain amount of lines is: (i) View invariant in 3D space, (ii) less complex compared to 3D surfaces, (iii) insensitive to colour and texture and (iv) the idea of curve recovery in 3D is supported by the growing progress of depth devices. Moreover, contours relate to the perspective of human behaviour since they are a crucial factor in the context of perception and cognition [Piz08].

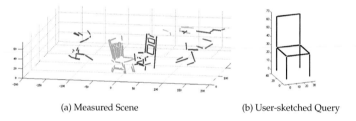

(a) Measured Scene   (b) User-sketched Query

Figure 5.3: The figure shows an exemplary scene consisting of two 3D objects represented by a set of 3D curve fragments. (**Left:**) The entire scene with two chairs placed in the centre surrounded by some clutter. The colours indicate the assignment of the 3D curves to the certain object clusters. The corresponding segmentation was obtained previously during the scene analysis. (**Right:**) The user-sketched chair query which is going to be used during the retrieval procedure. Please observe its highly abstracted character.

In analogy to [MYL13], the line or rather the curve fragment constitutes a major role in this project. Using lines by itself is a classical approach in the field of image processing. However, in 2D space lines are suffering from strong deformations caused by perspective changes. Being aware of this, the authors of [MYL13] propose to map this kind of representation into 3D space. Therefore, a depth image is acquired by employing a Kinect© sensor (cf. Section 2.1.1). Subsequently, the resulting depth data $\mathbf{I}^{(Depth)}$ is used in multiple contexts: First, the pure depth map is exploited for the detection of 2D edges. This is realised by utilising the Canny edge detector which is straightly applied to $\mathbf{I}^{(Depth)}$. Second, having these edges, they are *back-projected* into 3D with the goal of representing the object's geometry there. This recovery process or rather the back-projection, is implemented based on the point cloud $\mathcal{P}^3$ reconstructed from the data carried in $\mathbf{I}^{(Depth)}$ by involving both the intrinsic and extrinsic device parameters.

However, before the actual back-projection takes place, the 2D edges are combined to 2D fragments $C^2_{i=1,...,K}$ forming a curve that encompasses multiple edges. Afterwards, 2D-3D correspondences are established by assigning each pixel to a single point in $\mathcal{P}^3$. Strictly speaking, each 2D point is back-projected to its 3D counterpart by incorporating the depth information stored in $\mathbf{I}^{(Depth)}$. This, in turn, results in a set of 3D points $\mathcal{F}^3_{i=1,...,K}$ for each C^2_i. Finally, a 3D curve $C^3_{i=1,...,K}$ is iteratively approximated by generating 3D lines. This process is implemented by employing the RANSAC approach as introduced in Section 5.1. As a quick reminder, the line parameters are estimated by identifying the *consensus set* which minimises the error function while the amount of inliers is increased in respect of a given inlier standard deviation. The first fitting result is then added as a member to the final representation of C^3_i (arisen from C^2_i). By removing the *consensus set* from $\mathcal{F}^3_i$, a new subset $(\mathcal{F}^3_i)^C$ is retrieved only containing outlier elements. By re-invoking the RANSAC technique on $(\mathcal{F}^3_i)^C$, a second *consensus set* is detected and thus, a further line is added to the final representation. Consequently, even curved 2D fragments can be represented by this approach. The algorithm is repeatedly invoked until the amount of outlier drops below a user-defined threshold. Please notice that this iterative curve approximation is only applicable due to the working principle of the RANSAC technique, an LSM would fail in this special situation. Two fully recovered chair instances are exemplarily shown in Figure 5.3.

5.3 Feature Detection Using a Local Coordinate System

5.3.1 Estimation of an Intrinsic Object Coordinate System

The estimation of an object or local coordinate system is a challenging task due to the lack of missing information about both specific object characteristics and general *world knowledge*, e.g. a reference point. Having this in mind, the introduction of such a system appears to be even more important. Moreover, the set of unstructured 3D curves counteracts the aim of establishing an accurate and robust matching, whereas the LCS would drastically improve the performance of it in multiple aspects. In order to relax the sophisticated task of determining the LCS, two assumptions are made. First, all objects are assumed to be quadratic in the sense that their geometry can roughly be approximated by quadratically shaped primitives. Second, the objects are expected to be in their dedicated poses. In other words, the z-axis ($\dot{\mathbf{z}}$) is considered to be fixed during the entire matching. Particularly the latter is beneficial since it reduces the problem of estimating the LCS from 3D to 2D space.

The actual implementation operates on the cluster segments obtained by the scene analysis (Section 5.2.1). In more detail, all pixels covered by a cluster are passed as observation matrix to a Principal Component Analysis (PCA) approach. The result of the PCA delivers two axes, the so-called principle components indicating those dimensions having the highest variance. These axes are then interpreted as x-axis ($\dot{\mathbf{x}}$) and y-axis ($\dot{\mathbf{y}}$) of the LCS. Although this result does already deliver a good approximation in cases of perfectly detected objects, reality shows that deformations, occlusions and outliers drastically affect the desired result. In order to ensure the best alignment of these axes, a further refinement step has to be invoked based on the assumption that any better alignment can only appear in a range of $[-\pi/4, \pi/4]$ relating to the current location. Therefore, the suitability of the axes is checked by rotating them alternately in clockwise and counter-clockwise direction, e.g. $\alpha = [1°, -1°, 2°, -2°, \ldots]$. The idea behind this principle is simple: If the PCA already returned the optimal alignment, the fitness drops increasingly moving either clockwise or counter-clockwise. Thus, the algorithm can stop after a few iterations. If the PCA did not hit the optimal location, a better alignment exists in the surrounding area. Even in this case, the direction towards the wanted maximum can quickly be determined by closely observing the fitness value. Strictly speaking, any movement closer to the optimal configuration leads to an enhancement of this quantity. In the worst case the initialisation is placed at a local minimum between two valid solutions and thus, the entire interval has to be processed. Hence, the solution can be formulated as an optimisation problem in terms of maximising the fitness function:

$$\underset{\dot{\mathbf{x}}^{(\alpha)}}{\arg\max} \; f(\mathcal{P}^2, \dot{\mathbf{x}}^{(\alpha)}) \quad , \qquad (5.2)$$

where the fitness function $f(\cdot, \cdot)$ takes a set of 2D points depicted as $\mathcal{P}^2$, together with an α-rotated version ($\dot{\mathbf{x}}^{(\alpha)}$) of the initial x-axis ($\dot{\mathbf{x}}$). Please notice that $\mathcal{P}^2$ does not correspond to the previously used pixel set, this time the 2D points are generated based on the 3D curve segments which are representing the object. Therefore, the start and end point of each line are projected to the $\dot{\mathbf{x}}/\dot{\mathbf{y}}$-plane of the object's space forming a further set of 2D points. Figure 5.4 illustrates this projection as well as the new point set which is defined by the blue circles in both images.

Moreover, it is worth mentioning that the axis with the highest eigenvalue does supply

(a)   (b)

Figure 5.4: The figure shows an exemplary instance of $\mathcal{P}^2$ defined by the blue circles. The point set is obtained by projection the start and end point of all 3D lines to the *bottom* or rather the $\dot{x}/\dot{y}$-plane of the object's space. **(Left:)** The 3D chair in its dedicated pose, standing on the $\dot{x}/\dot{y}$-plane. **(Right:)** Top-View on the same object instance as illustrated left.

the x-axis. Given these inputs, the function returns a scalar indicating the degree of symmetry between the disjoint subsets $\mathcal{A}^{(\alpha)} \subset \mathcal{P}^2$ and $\mathcal{B}^{(\alpha)} \subset \mathcal{P}^2$ which arise by slicing $\mathcal{P}^2$ along $\dot{x}^{(\alpha)}$ crossing the centre of gravity (COG):

$$\forall \alpha \quad \mathcal{A}^{(\alpha)} \cap \mathcal{B}^{(\alpha)} = \emptyset \quad \text{and} \quad \mathcal{A}^{(\alpha)} \cup \mathcal{B}^{(\alpha)} = \mathcal{P}^2 \quad . \tag{5.3}$$

This is demonstrated in Figure 5.5a showing a good approximation of $\dot{x}$ (green line). Here the axis starts at the COG (red dot) and clearly separates $\mathcal{P}^2$ into two subsets indicated by the red and the blue circles. Afterwards, all points in $\mathcal{A}^{(\alpha)}$ and $\mathcal{B}^{(\alpha)}$ are further projected on $\dot{x}^{(\alpha)}$, respectively. By traversing all projections according to their position on $\dot{x}^{(\alpha)}$, a natural order emerges in each subset (cf. Figure 5.5a). Finally, the *point-to-axis* distance is calculated defined as the length of the vector connecting the point with its projection on the axis (dotted lines in Figure 5.5a).

Finally, a time series matching technique (e.g. DTW or EMD) is utilised to calculate the actual degree of symmetry between $\mathcal{A}^{(\alpha)}$ and $\mathcal{B}^{(\alpha)}$. Please recognise that the EMD is capable of incorporating the $\dot{z}$ value as additional weight in combination to the pure ground distance matrix. The actual alignment is then found by detecting that $\dot{x}^{(\alpha)}$ leading to the lowest symmetry distance. Having this axis, the second one can easily be derived. If $\dot{x}^{(\alpha)}$ does not point to the direction which is accommodating the higher amount of points (e.g. to the back of a 3D chair), the method switches the caption of the axes. If necessary, it also negates the axis orientation to fulfil this condition.

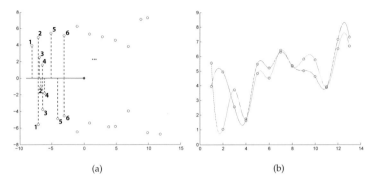

(a)   (b)

Figure 5.5: The figure illustrates the working principle of the refinement step in order to obtain the best fitting object coordinate system. **(Left:)** Top-view on the $\dot{x}/\dot{y}$-plane showing the start and end points of all lines (circles). Moreover, a good approximation of $\dot{x}$ (green line) is given that separates the points of $\mathcal{P}^2$ into two sets, namely $\mathcal{A}^{(\alpha)}$ and $\mathcal{B}^{(\alpha)}$ indicated with the colours red and blue. Please keep in mind that all points are placed inside the world coordinate system. Besides the ordering scheme derived by the point projections on the axis, the *point-to-axis* distance is visualised by the dotted lines. **(Right:)** An illustrative plot of these *point-to-axis* distances (vertical axis) which are drawn according to their corresponding point order (horizontal axis) in both subsets.

5.3.2 Detection of Object Characterising Feature Points

After establishing a local coordinate system, the chosen object representation must be mapped to a PSSGM friendly structure in order to adopt its concept to this project task. In other words, the idea of using shortest paths as introduced in [BL08] has to be applied to the unstructured set of 3D curve segments. Consequently, feature points are required which can be exploited in analogy to the skeleton end points. Therefore, the feature space is restricted to the start and end points of all 3D line fragments $(\mathcal{P}^3)^{\mathsf{C}}$. In order to further alleviate this problem, the same assumptions concerning the object's pose and geometry are reused from the previous section. In the following, a robust procedure is proposed capable of localising significant visual and structural regions of the 3D shape which are subsequently employed for the generation of shortest paths. Therefore, the method starts with the detection of an initial set of features having a highly generic character in order to stay compatible to a wide range of objects. The actual set is then determined among a sequence of tuples $(\mathbf{p}, \mathbf{q})_i$ having the highest distance to each other compared to all remaining ones:

$$\mathbb{D} = \{(\mathbf{q}, \mathbf{p})_i \mid \mathbf{q}, \mathbf{p} \in (\mathcal{P}^3)^C : \mathbf{q} \neq \mathbf{p} \quad \wedge \quad \forall \hat{\mathbf{q}} \in (\mathcal{P}^3)^C \backslash \{\mathbf{q}, \mathbf{p}\} \to |\mathbf{q} - \mathbf{p}| > |\hat{\mathbf{q}} - \mathbf{p}|\} \quad . \tag{5.4}$$

By ordering this sequence according to the distance values, the first four positions (or two if the tuples are not allowed to be symmetric) are considered as initial feature points which are referred to as $(\mathbb{D}^C)^{(0)}$ in the following. If the object has been detected robustly, the tuple in $(\mathbb{D}^C)^{(0)}$ should only encompass four points in total which create two imaginary diagonals as illustrated in Figure 5.6a. While the approach performs well in the absence of occlusions or heavy outliers, the approach strongly suffers from their existence. In order to tackle this problem and to suppress their influence, an additional constraint has to be fulfilled. Verbally expressed, each tuple in $(\mathbb{D}^C)^{(0)}$ has to maintain a 90 degree angle relation to the remaining points in $(\mathcal{P}^3)^C$. A tuple passes this condition if the stated angle relation can be established at least once:

$$\{(\mathbf{q}, \mathbf{p}) \mid \exists \hat{\mathbf{q}} \in (\mathcal{P}^3)^C : f(\mathbf{q}, \mathbf{p}, \hat{\mathbf{q}}) = 1\} \quad , \tag{5.5}$$

where

$$f(\mathbf{q}, \mathbf{p}, \hat{\mathbf{q}}) = \begin{cases} 1, & \text{if} \quad |\frac{\pi}{2} - \arccos(\langle \mathbf{p} - \hat{\mathbf{q}}, \mathbf{q} - \hat{\mathbf{q}} \rangle)| < \mu^{(\text{angle})} \\ 0, & \text{otherwise} \end{cases} \quad . \tag{5.6}$$

Please notice that $\langle \cdot \rangle$ depicts the dot product and $\mu^{(\text{angle})}$ is a configurable parameter steering the 90° angle deviation tolerated by the proposed approach. Afterwards, a non-maximum suppression (NMS) is performed as explained later in this section. For the time being, it is assumed that $\|(\mathbb{D}^C)^{(0)}\| \geq 2$ in order to detect further feature points as described below. The basic concept of this analysis determines new features iteratively with respect to the already existing ones. Therefore, the points in $(\mathcal{P}^3)^C \backslash (\mathbb{D}^C)^{(0)}$ are successively selected in order to verify their state towards being a member of the feature point set. For this purpose, two points $\mathbf{p} = (p_1, p_2, p_3)^\mathsf{T}$ and $\mathbf{q} = (q_1, q_2, q_3)^\mathsf{T}$ are taken from $\mathbb{D}^C$, whose coordinates significantly differ in two dimensions while they remain similar in one:

$$\begin{aligned} (p_1 - q_1)(p_2 - q_2)(p_3 - q_3) &\approx 0 \wedge \\ |(p_1 - q_1)(p_2 - q_2) + (p_1 - q_1)(p_3 - q_3) + (p_2 - q_2)(p_3 - q_3)| &\gg 0 \quad . \end{aligned} \tag{5.7}$$

Please notice that this procedure can only be realised due to the LCS estimated previously. By swapping one of the coordinates analysed as dissimilar, two virtual points are generated, e.g. $\mathbf{p}^\circ = (q_1, p_2, p_3)^\mathsf{T}$ and $\mathbf{q}^\circ = (p_1, q_2, q_3)^\mathsf{T}$. Subsequently, $\mathbf{p}^\circ$ and $\mathbf{q}^\circ$ are given to a *Nearest*

Neighbour approach like a k-dimensional tree structure (k-d tree) with the intent to find their closest point correspondences in $(\mathcal{P}^3)^\complement$. To reduce the amount of false positives which might appear during that procedure, the condition in (5.6) is reused to accept either $\mathbf{p}^\circ$ or $\mathbf{q}^\circ$ as new feature points, e.g. $(\mathbb{D}^\complement)^{(1)} = (\mathbb{D}^\complement)^{(0)} \cup \{\mathbf{p}^\circ\}$. This procedure is repeated until all points in $(\mathbb{D}^\complement)^{(0)}$ have been processed and terminates with a further execution of the NMS approach. If new feature points have been found in the last cycle the process is re-invoked with $(\mathbb{D}^\complement)^{(1)}$. Figure 5.6a illustrates this based on the exemplary instance of a chair where the virtual points are drawn in cyan and their closest neighbours in red. Finally, the expected feature set should encompass the outer shape of an object, e.g. six points for a chair, namely its legs and the upper points of its back (green and pink dots in Figure 5.6c).

In order to detect intermediate object layers, like the seat of a chair, the working principle of the previous approach has to be extended in the following. Starting with $(\mathbb{D}^\complement)^t$, the current feature point set, a further detection run is performed, but this time two features $\mathbf{p},\mathbf{q}$ and *two* points $\mathbf{p}^\star,\mathbf{q}^\star$ of $(\mathcal{P}^3)^\complement \setminus (\mathbb{D}^\complement)^t$ are taken into consideration. The points $\mathbf{p}^\star,\mathbf{q}^\star$ are added to $(\mathbb{D}^\complement)^{t+1}$ if they form a rectangular-related primitive as illustrated in Figure 5.6c with a tolerated distortion. As shown there, this fitting procedure leads to an over-representation of the seat corners in terms of feature points. The solution for this is again the non-maximum suppression technique as applied previously and which shall be introduced now. The procedure consists of two cycles: First, the features are clustered along the $\mathbf{z}$-axis ($\dot{\mathbf{z}}$). Second, each cluster is projected on the $\dot{\mathbf{x}}/\dot{\mathbf{y}}$-plane of the object's LCS. Afterwards, each projection is assigned to that quadrant which accommodates it:

$$f(\mathbf{p}') = \begin{cases} 1, & \text{if} & 0 \leq \psi(p'_2, p'_1) < \frac{\pi}{2} \\ 2, & \text{if} & \frac{\pi}{2} \leq \psi(p'_2, p'_1) < \pi \\ 3, & \text{if} & 0 > \psi(p'_2, p'_1) < -\frac{\pi}{2} \\ 4, & \text{if} & -\frac{\pi}{2} \geq \psi(p'_2, p'_1) \geq -\pi \end{cases}, \tag{5.8}$$

where $\mathbf{p}' = (p'_x, p'_y)^\mathsf{T}$ corresponds to $\mathbf{p} = (p_x, p_y, p_z)^\mathsf{T}$. Moreover, the function $\psi(\cdot, \cdot)$ is defined as: $2 \cdot \arctan(\mathbf{e}_y(\sqrt{\mathbf{e}_x^2 + \mathbf{e}_y^2} + \mathbf{e}_x)^{-1})$ and better known as *atan2* from the field of computer programming languages. Using this function, a sign-attached outcome is retrieved which easily allows the assignment of a point to one of the quadrants. According to Figure 5.6d, three quadrants encompass two features, respectively and one only a single point. In order to ensure that each quadrant is only represented by one feature, the point with the highest distance to the object's COG is selected as representative for it.

The implementation of $\dot{\mathbf{z}}$ clustering procedure first determines the maximum and minimum $\dot{\mathbf{z}}$ coordinate of all features ($\dot{\mathbf{z}}^{\max}, \dot{\mathbf{z}}^{\min}$). Subsequently, the final feature set $(\mathbb{D}^\complement)^{(\text{final})}$ is

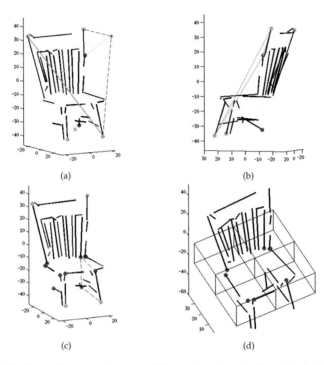

(a)   (b)

(c)   (d)

Figure 5.6: The figure illustrates the proposed feature detection pipeline. **(Top Left:)** Initial feature points (green dots) and two virtual ones (cyan) coupled with their closest point correspondences (red). Please also observe the imaginary diagonal (orange) used to determine the initial point set. **(Top Right:)** False positive removal by enforcing the condition of Equation (5.6) (upper red dot). **(Bottom Left:)** Valid feature set describing the outer shape as well as the intermediate layer (blue dots). **(Bottom Right:)** 2D clustering scheme performing the NMS as demonstrated on the example of the intermediate features. The removal depends on the distance to the object's COG.

separated into two subsets by assigning each feature point either to $\dot{z}^{max}$ or $\dot{z}^{min}$ according to their distance:

$$g(\mathbf{p}) = \begin{cases} \dot{z}^{max}, & \text{if} \quad |\mathbf{p}_3 - \dot{z}^{max}| < |\mathbf{p}_3 - \dot{z}^{min}| \\ \dot{z}^{min}, & \text{if} \quad |\mathbf{p}_3 - \dot{z}^{min}| < |\mathbf{p}_3 - \dot{z}^{max}| \end{cases}, \quad (5.9)$$

with $g(\cdot)$ being the assignment function. However, this separation approach does only work appropriately for stands and tables. For chair instances a third level is required, defined as: $\dot{z}^{mid} = (\dot{z}^{max} - \dot{z}^{min})/2$. Section 5.6 describes a more complex approach to estimate the needed number of separation levels automatically.

5.4 Shortest Path Computation Based on a Modified Dijkstra Implementation

After introducing the concept of feature points and their analogy to skeleton end points in the previous section, this content part is devoted to the extraction of shortest paths based on 3D curves. Although it might sound trivial to compute paths on a set of lines representing the object's geometry with the intent to establish a connection between two feature points, a closer look reveals a higher complexity than expected. This observation mainly correlates to multiple aspects, e.g. the lines are (i) unstructured, (ii) not complete, (iii) different in their number and (v) not connected at all. Moreover, the object representation suffers from outliers as well as measurement inaccuracies. Thus, the native application of a shortest path algorithm would fail without pre-processing this set of curve fragments.

The first attempt which has been investigated to solve this problem introduces a cost matrix $\mathbf{C}'$ carrying the weights (Euclidean distance) to connect all start and end points of all lines with each other. Therefore, two cases have to be distinguished: The measured 3D curves on the one hand and *virtual* connections between them on the other. Finally, $\mathbf{C}'$ is given to a modified Dijkstra implementation determining a path which alternates between these two curve segments as illustrated in Figure 5.7a.

Moreover, the path is restricted to starting and and ending with a measured line. Although the procedure leads to promising results, it struggles in situations which suffer from a higher amount of outliers. Instead of tracing the object's visual structure, the path is *jumping* over outlier curves to reach its destination as demonstrated in Figure 5.7b. This is a valid behaviour from the methodological view in order to find the shortest path even though it is not desired. Figure 5.7b clearly shows this disadvantage towards the aim to reflect the specific characteristics of the object's geometry. To overcome this *jumping phenomena* a

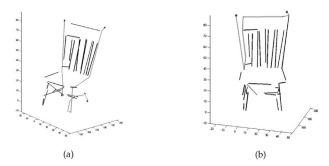

(a)   (b)

Figure 5.7: The figure shows some problematic cases in context of the alternating path tracking mechanism. **(Left:)** Comparison of two paths established with the aim to model the same geometrical relation, namely from the upper back down to the leg (red dots). On the right side, this path has perfectly been established, while the opposite side did not perform well. This is caused by the almost missing leg on the left side of the chair. Thus, the algorithm did not find an appropriate connection with the result of an entirely misled path. **(Right:)** Since the algorithm is enforced to start with a measured curve, a wrong direction is taken with the effect of a strong deformation at the beginning. Both situations are highly problematic and strongly affect the robustness of this approach.

nearest neighbour approach controls the neighbourhood of each point with the intention of connecting only those curves whose distances are below a certain threshold. Please notice that this constraint is only enforced for virtual connections, the distance between two points sharing the same curve segment is not limited by this approach. Roughly speaking, a sphere with radius $\mu^{(\text{radius})}$ is spanned around a point $\mathbf{x}^\star \in (\mathcal{P}^3)^\mathsf{C}$ in order to gather all other points $\mathbf{x}_i^\circ$ inside this sphere to constitute the set of valid matching partners:

$$\{\, \mathbf{x}^\circ \mid \mathbf{x}^\star, \mathbf{x}^\circ \in (\mathcal{P}^3)^\mathsf{C} : \mathbf{x}^\star \neq \mathbf{x}^\circ \wedge \mid \mathbf{x}^\star - \mathbf{x}^\circ \mid < \mu^{(\text{radius})} \,\} \quad . \tag{5.10}$$

However, this solution did not solve the problem so far. Finding an appropriate value for $\mu^{(\text{radius})}$ is challenging due to the quality of $\mathcal{P}^3$ in terms of accuracy, completeness and false positives. The fitness of $\mathcal{P}^3$ strongly depends on multiple parameters, e.g. the scene light, perspective or occlusion. Hence, the curves are not equally distributed or fully recovered. Consequently, the distances between the curves are not reliable, even among instances of the same object class. In addition to this, observations revealed that the enforcement of starting with an existing curve segment does often fail or leads to strong deformations as illustrated in Figure 5.7b. As a result of this, the tracing mechanism cannot be established on top of this approach. However, in the following another concept is proposed which behaves more insensitive towards these obstacles.

5.4.1 Geometry-driven Establishment of Feature Networks

In order to remain in the field of *computational geometry*, a second and more robust approach will be introduced in the following content. As already discussed above, the explicit use of 3D curves for establishing shortest paths leads to unsatisfying results. Thus, the second method exploits the curves only implicitly to establish a network between the features. Therefore, the curve fragments are considered as a kind of evidence confirming the existence of a justified connection between two feature points. The objective of this routine is to create a network of virtual links between all features according to the evidence derived from the geometry or rather from the set of 3D curves which are representing the object.

Virtual Link   A non-existing curve segment virtually spanned between two features. The benefit of using only virtual lines or rather a network of them is their binary character which is not affected by outliers which typically have the potential to corrupt the path tracking. The decision whether or not a connection exists is made on top of an evidence-driven implementation. Consequently, the task of determining shortest paths can be increased drastically in terms of robustness and accuracy and thus, the overall matching result.

A virtual link between two features is established by collecting evidence about the existence of a real structural connection by taking the entire set of 3D curves into account. Therefore, a set of rules is utilised to check whether there exists at least one measured curve segment which satisfies the constraints of being a structural link inside the object's geometry:

Curve Length   The length of a measured curve fragment has to be greater than $\mu^{(A)}$ in order to be considered as a possible candidate.

Length Ratio   The ratio between the length of an existing 3D curve segment a and the virtual link b which is to be confirmed by it has to be greater than $\mu^{(R)}$: $a/b > \mu^{(R)}$.

Spatial Arrangement   The spatial arrangement of both the curve a and the link b has to be similar. Strictly speaking, the start $\mathbf{a}^{(s)}$ or the end point $\mathbf{a}^{(e)}$ of a has to be closely located to one of the feature points $\mathbf{p}$ or $\mathbf{q}$ to be connected inside the network:

$$\| \mathbf{a}^{(s)} - \mathbf{p} \| \leq \mu^{(C)} \vee \| \mathbf{a}^{(s)} - \mathbf{q} \| \leq \mu^{(C)} \vee \| \mathbf{a}^{(e)} - \mathbf{p} \| \leq \mu^{(C)} \vee \| \mathbf{a}^{(e)} - \mathbf{q} \| \leq \mu^{(C)} \quad , \qquad (5.11)$$

where $\mu^{(C)}$ determines the tolerated distance.

Orientation This rule connects to the spatial arrangement which additionally ensures an almost parallel orientation of a to b. The tolerated deviation is controlled by the parameter $\mu^{(W)}$ which indicates the upper limit of the opening angle between both segments. Assuming that $\| \mathbf{a}^{(s)} - \mathbf{p} \|$ returns the closest spatial arrangement, the opening angle is calculated as:

$$\arccos(\langle \frac{\mathbf{a}^{(e)} - \mathbf{a}^{(s)})}{\|\mathbf{a}^{(e)}\| \cdot \|\mathbf{a}^{(s)}\|}, \frac{\mathbf{q} - \mathbf{p}}{\|\mathbf{q}\| \cdot \|\mathbf{p}\|} \rangle) \leq \mu^{(W)} \quad , \tag{5.12}$$

where $\langle \cdot \rangle$ depicts the dot product.

If all constraints are fulfilled, the observed feature pair $(\mathbf{p}, \mathbf{q})$ is marked as virtually connected. An exemplary result is illustrated in Figure 5.8, where all connections have been established correctly. The coloured lines inside the figure indicate the chosen curve segments which have been detected as evidence for the existence of structural connections between the corresponding feature pairs. Please keep in mind that *only* the object's geometry has been exploited without including any further semantic information.

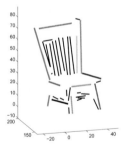

Figure 5.8: The figure illustrates the principal of determining possible connections between the detected features. These associations are crucial for the success of robustly localising shortest paths. The coloured lines indicate the curve fragments which confirm the establishment of a virtual link between two feature points. In this example all features have been connected correctly.

It is worth knowing that the great advantage of this approach lies in the absence of using explicitly existing 3D curve fragments during the estimation of feature connections. By limiting the algorithm domain to the network of virtual links which have been established implicitly as straight lines, any outlier jumping is entirely suppressed during the computation of shortest paths. Thus, the actual outcome is more robust and accurate compared to the previous approach. Additionally, the generic character of the method increases the variety of objects which can be processed with it. Being aware that the network might suffer from a certain amount of wrong connections, some modifications to the Dijkstra algorithm will be introduced next. These adaptations alleviate the negative influence of these false positives.

5.4.2   Extended Dijkstra Implementation Operating on Feature Networks

After creating the foundation to compute shortest paths between the features of an object, this section introduces adaptations of the original Dijkstra algorithm which are required to utilise the information given by the feature network. These modifications are necessary due to the nature of symmetry embodied by many real-world objects. In consequence of this, it might occur that the shapes of two paths are different even though they are established between the same feature point pair with the expectation to represent the same geometrical structure. The user-generated query instance of a chair constitutes a good example for this issue. Figure 5.9 demonstrates the problem based on the path from the lower left to the upper right of the chair. Please observe the different shapes of both paths connecting the same feature constellation, respectively. The actual problem arises due to the fact that the paths are of the same length and cannot be distinguished uniquely. Hence, an appropriate selection criterion has to be defined which supports the algorithm in its decision of which path to take.

Other kinds of path irritations induced by measurement errors during the depth acquisition or the 3D curve recovery process, e.g. the curves exhibit displacement errors or are not recognised at all. Consequently, the method runs the risk of returning an unexpected result which increases the overall path distance and thus, the corresponding matching costs. In order to tackle this problem as well as the symmetry issue stated above, two additional mechanisms are proposed to extend the functionality of the Dijkstra algorithm concerning the demands of the input data: First, the algorithm is enabled to monitor all shortest paths of the query. Second, the method is designed to take this documentation with the intent to find similar paths inside the target object.

For the purpose of documenting the path's shape, a simple *route tracking* approach is exploited which captures the route changes of a path. Since the objects are in 3D, three directional variations are recorded in terms of *left/right*, *up/down* and *front/back*. Therefore, the LCS (cf. Section 5.3.1) is reused and a path definition matrix $\mathbf{P}_{N' \times 3}$ is defined to represent the query paths:

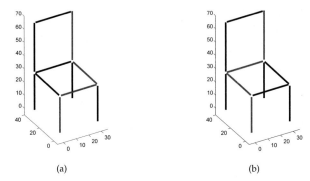

(a)   (b)

Figure 5.9: The figure illustrates the problem of path ambiguities caused by the natural symmetry of real-world objects. This concrete example utilises the user-generated query chair to demonstrate this issue. Please observe the different shape of both paths (drawn in red), even though they are connecting the same pair of features. Thus, the aim to describe similar geometrical relations by almost identical paths failed in this example. Moreover, both paths are of the same length.

$$
\mathbf{P}_{N'\times 3} = \begin{pmatrix} u_{1,1} & u_{1,2} & u_{1,3} \\ \vdots & \vdots & \vdots \\ u_{i,1} & u_{i,2} & u_{i,3} \\ \vdots & \vdots & \vdots \\ u_{N',1} & u_{N',2} & u_{N',3} \end{pmatrix}, \tag{5.13}
$$

with

$$
\begin{aligned} u_{i,j} &\in \{-1,0,1\} \\ \sum_{i=1}^{3} |u_{i,j}| &= 1 \end{aligned} \tag{5.14}
$$

Please notice that the value of N' corresponds to the number of course breaks, while the columns indicate the dimension. Hence, each row always encodes one directional variation at a time. Consequently, if a path changes its orientation five times, the size of $\mathbf{P}$ equals 5×3. Moreover, if the path starts going upwards, the first row is populated with $(0,0,1)$ if it was going left it would be $(-1,0,0)$ and so on. At the end, each path is assigned to its path definition matrix $\mathbf{P}_i$. Subsequently, this $\mathbf{P}_i$ is given to the extended Dijkstra algorithm as additional input together with the target object to localise a similar path as in the query. This mechanism is substantial to retrieve stable path correspondences.

Moreover, the lack of information, e.g. caused by errors or occlusions during the curve recovery process, can be compensated slightly by the path definition matrix. Strictly speaking, the algorithm is allowed to establish a virtual link if it is required by the data provided in **P**. In more detail, the algorithm selects the closest feature point (with the smallest opening angle) located in that direction which is proposed by the path definition matrix but not established in the feature network. If this new connection does not support any further progress of finding a proper path, the algorithm discards it from the matching. Figure 5.10 shows the same path on both the user-generated query (5.10a) and a measured target object (5.10b).

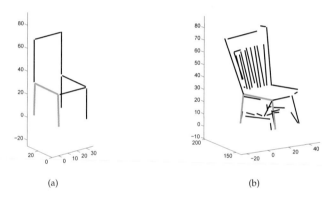

(a)   (b)

Figure 5.10: The figure shows an exemplary shortest path established based on the virtual link network and the extended Dijkstra algorithm. **(Left:)** Calculated on the user-generated query. **(Right:)** Localised on the measured target object by utilising the path definition matrix.

5.5 Path Descriptor Based on Relative Angles

At this stage of the processing pipeline, the following progress can be summarised: (i) The raw input data has been converted to a 3D curve representation, (ii) a user-generated query object has been created, (iii) feature points have been located and a feature network has been established. Moreover, by taking this information coupled with an extended version of the Dijkstra algorithm, shortest paths have been retrieved for any feature combination in both the query and the target object.

Having this in mind, this section now introduces an appropriate description to encode the characteristic shape of these previously obtained paths. While such a descriptor is necessary to accomplish the actual matching goal, its quality drastically contributes to the success or failure of the system (Section 5.6). The description by itself is used to calculate the matching costs of two features, respectively. Consequently, the path descriptors $\hat{x}_{i=1,\dots,K}$ carry meaningful values specifying a certain path in order to retrieve reliable matching data. Furthermore, the descriptors should be RST invariant. For this purpose, *relative angles relations* are proposed which can be derived in cooperation with the LCS. In more detail, the implementation encompasses two angles α and β which are calculated at equally distributed points along the shortest path as already introduced in [BL08]. The actual point definition is illustrated in Figure 5.11 where a 3D point is uniquely defined by two angles whose values are straightly derived from the object's coordinate system.

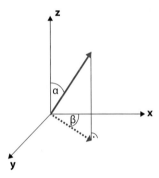

Figure 5.11: The figure demonstrates the concept behind the proposed path descriptor. Two angles α and β are calculated for multiple points which are uniformly distributed along the shortest path. The result of this operation is a sequence of RST invariant tuples describing the specific shape of a certain path in relation to its start point and the local coordinate system.

In analogy to [BL08], the procedure scans the path structure with a set of M equidistantly distributed points as illustrated in Figure 5.12. The procedure returns a sequence $\hat{x}$ consisting of M tuples which encode the relative position of each sample $e_{j=1,\dots,M} \in \mathbb{R}^3$ in relation to a reference point $e_0 \in \mathbb{R}^3$. In context of this project, e_0 refers to the feature point from which the corresponding path emanates. In order to calculate the angle values, the sample points have to be extended to *sample vectors* $e^{\star}_{j=1,\dots,M}$ connecting each $e_{j=1,\dots,M}$ with its reference e_0.

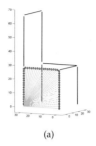
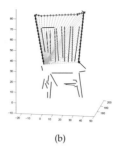

(a)   (b)

Figure 5.12: The two illustrations demonstrate the principle behind the proposed path description. The red circles indicate the position of a single sample point $\mathbf{e}_j$ and the lines drawn in cyan are the sample vectors $\mathbf{e}_j^{\star}$ connecting $\mathbf{e}_j$ with $\mathbf{e}_0$ the starting point of the currently observed path.

$$\mathbf{e}_j^{\star} = \mathbf{e}_j - \mathbf{e}_0 \quad . \tag{5.15}$$

Taking these sample vectors, the angle α is calculated as:

$$\alpha = \arccos(\langle \mathbf{e}_j^{\circ}, \dot{\mathbf{z}} \rangle) \quad , \tag{5.16}$$

with $\mathbf{e}_j^{\circ} = \mathbf{e}_j^{\star}/\|\mathbf{e}_j^{\star}\|$ and $\dot{\mathbf{z}}$ being the z-axis of the LCS. The dot product is depicted by the symbol $\langle \cdot \rangle$. The computation of β works similar to the calculation of α, where $\mathbf{e}_j^{\circ}$ is first projected on the $\dot{\mathbf{x}}/\dot{\mathbf{y}}$-plane:

$$\beta = \arccos(\langle \mathbf{e}_j^{\bullet}, \dot{\mathbf{x}} \rangle) \quad , \tag{5.17}$$

with $\mathbf{e}^{\bullet} = [\langle \mathbf{e}_j^{\circ}, \dot{\mathbf{x}} \rangle, \langle \mathbf{e}_j^{\circ}, \dot{\mathbf{y}} \rangle, 0]^{\top}$ and $\dot{\mathbf{x}} / \dot{\mathbf{y}}$ being the x-axis ($\dot{\mathbf{x}}$) and y-axis ($\dot{\mathbf{y}}$), respectively. The angles α and β are then stored in form of a tuple inside the path representation vector $\hat{\mathbf{x}}$. The length of the vector is equivalent to the number of points which have been exploited for the sampling, namely $\|\hat{\mathbf{x}}\| = M$:

$$\hat{\mathbf{x}} = [(\alpha, \beta)_1, \ldots, (\alpha, \beta)_M]^{\top} \quad . \tag{5.18}$$

For the purpose of 3D object recognition, each path has to be encoded based on this descriptor. Subsequently, the vectors are employed during the matching process in Section 5.6, where the dissimilarities between two paths is used to determine the matching costs of two features. This indicator is crucial as it constitutes the foundation for the final establishment of correspondences and thus, for the overall dissimilarity between the query and target. The actual path distance value is estimated based on a time series matching approach like the DTW or the EMD. Therefore, $\hat{x}$ is separated into two disjoint sub-sequences, namely $\hat{x}^{(\alpha)}$ and $\hat{x}^{(\beta)}$, which are respectively carrying the α and the β elements of the original vector. These sub-sequences are then given to the distance measurement approach where they are processed independently from each other. Like in Section 5.3.1, the EMD is able to take the length of each sample vector ($\|e_j^\star\|$). On the one hand, this additional information facilitates the discrimination of each path. On the other hand, the approach loses its property of being RST invariant by incorporating this further data dimension.

5.6 3D Object Retrieval Based on 3D User-Sketched Queries

The final section of this project is dedicated to the actual matching procedure. In the following, a detailed overview of each component of the 3D object retrieval system is given, needed to implement a robust and accurate recognition of 3D objects. One key feature of the system is its capability of accepting user-generated queries as input. Especially this user-interaction step appreciates a high acceptance due to the service of low level primitives (the 3D curve fragments) since it drastically reduces the complexity during the sketching process of the object.

The only stage which has not been considered in detail yet, is the computation of the overall similarity value between the query G and the target G'. All information which is necessary to perform this step is available at this point and only the following activities are pending to apply the matching in analogy to the PSSGM: (i) An ordering scheme has to be proposed to order the features deterministically, (ii) the matching costs ($c(v_i, u_j)$) have to be calculated for each feature pair $v_{i=1,...,\|G\|} \in G$ and $u_{j=1,...,\|G'\|} \in G'$ and (iii) the actual object matching procedure has to be performed. Finally, a correspondence configuration is retrieved with minimum costs. Please notice that this matching principle is still the one which was presented in context of the PSSGM (cf. Section 3.2), but this time the method operates on depth sensory data and a completely different representation type.

5.6.1 Feature Point Ordering Scheme

A highly crucial requirement for the following matching concept is a reproducible and deterministic feature ordering scheme. The reason for this is attributed to the Optimal Subsequence Bijection (OSB) or rather to its working principle (cf. Section 2.4.3.2). As a quick reminder, the OSB is used to estimate the costs for assigning two features during the matching. In more detail, the OSB calculates these costs based on two time series and thus, the method becomes sensitive to the element order in theses sequences. Assuming that the elements were randomly placed, the OSB would not be able to return a meaningful result. Although the problem sounds trivial and can easily be solved in 2D, it is more challenging to find an equivalent solution in 3D. Simple approaches like *distance-to-ground* or *distance-to-camera* are not appropriate due to a changing perspective and noisy input data.

In this project, another solution is presented which incorporates the object's LCS. With access to this reference system, the entire ordering scheme can operate on top of it. Altogether, the proposed mechanism is designed to be highly generic with the intent to be applicable for a wide range of object classes. Therefore, the algorithm starts to determine an appropriate number of disjoint areas along the z-axis ($\dot{z}$). Please remember, the object is assumed to be in its dedicated pose. Hence, only the z-coordinates $p^{(z)}$ of all features $\mathbf{p}_i = (p_i^{(x)}, p_i^{(y)}, p_i^{(z)})^\top$ have to be extracted and subsequently arranged in descending order inside a vector $\mathbf{v}$. By subtracting adjacent elements in $\mathbf{v}$, a new vector $\mathbf{v}^\star = (v_0 - v_1, v_1 - v_2, \dots, v_{\|\mathbf{v}\|-2} - v_{\|\mathbf{v}\|-1})^\top$ is obtained which can be passed to a peak detection method analysing the magnitudes of these $\dot{z}$-differences (cf. Figure 5.13a). A peak is detected if its scale is above a certain threshold, e.g. the mean value or the standard deviation of $\mathbf{v}^\star$.

These peaks are back-projected to their corresponding z-values which are going to be exploited for the detection of the feature clusters along the z-dimension. Recalling that the features are segmented according to their distance towards a certain reference point residing on the $\dot{z}$, a closer look at Figure 5.13a discovers that these references are located between two peaks, respectively. In order to calculate these z-coordinates, the peaks are retraced to their differences which, in turn, grant access to the values in $\mathbf{v}$ responsible for their arising. Once the values are extracted in terms of $[p_i^{(z)}, p_j^{(z)}, p_m^{(z)}, p_n^{(z)}]^\top$ with $p_i^{(z)} \leq p_j^{(z)} \leq p_m^{(z)} \leq p_n^{(z)}$ and $i \neq j \neq m \neq n$, they can easily be employed to determine the desired z-coordinate of the cluster (or valley): $\dot{z}^{(\text{cluster})} = p_j^{(z)} + p_m^{(z)}/2$. After processing all valleys according to the instructions given above, the minimum and the maximum z-value of all features located in v_0 and $v_{\|\mathbf{v}\|-1}$ are marked as additional cluster references. Finally, each element in $\mathbf{v}$ is respectively assigned to its closest clusters based on its minimum distance to one of the z-references (cf. Section 5.3.2). The actual order is then accomplished as follows: Firstly, all members in each cluster are projected independently on disjoint $\dot{x}/\dot{y}$-planes. Secondly, an arbitrary feature is selected as sequence start and labelled with the index zero: $\mathbf{p}_{i,0}$ (red dot in Figure 5.13b). Thirdly, the cluster members of the active point are ordered clockwise with respect to $\mathbf{p}_{i,0}$. Fourthly, in order to bridge the process to the other points, $\mathbf{p}_{i,0}$ is projected into the next upper cluster (pink dot). Fifthly, the algorithm continues the ordering process by selecting the closest point to the projection and by traversing the remaining ones in

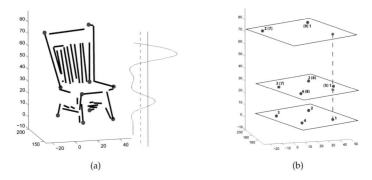

(a)   (b)

Figure 5.13: The figure illustrates the proposed ordering concept regarding the feature points which have been determined for the object (blue dots). **(Left:)** A measured chair, its features and the deviation-plot of the successively arranged z-values. The green line indicates the magnitude of these differences, while the mean value and the standard deviation are illustrated by the dashed and the blue line, respectively. One can easily observe that both quantities are qualified for the purpose of detecting peaks. **(Right:)** Abstracted representation of three clusters and their members corresponding with the features on the left side. Moreover, the cluster members are projected on disjoint $\dot{x}/\dot{y}$-planes. By selecting one feature as starting point (red dot), the remaining ones are ordered in clockwise direction to it. The connection between the clusters is then established by projecting the start point (red dot) into the remaining clusters (pink dot). There the method operates in the same manner as before by continuing the counting process with the closest feature to this projection.

accordance to the principle stated above. Please notice that the procedure automatically jumps from the highest to the lowest cluster if no further area exists above the current one. The technique terminates when all clusters have been processed and finally merged into one sequence according to the just now identified point order, e.g. $[\mathbf{p}_{i,0}, \mathbf{p}_{i,1}, \mathbf{p}_{i,2}, \cdots, \mathbf{p}_{\|\mathbf{v}\|-1}]^\top$. The whole procedure is demonstrated in Figure 5.13b, where the features are drawn as blue and the projections as pink dots.

5.6.2   Maximum Weight Cliques

This section is dedicated to the actual 3D object matching and recognition task. Please remember that the working principle of this project is designed in analogy to [BL08]. Nevertheless, the method differs in multiple aspects from the one proposed by Bai et al. The most obvious difference is the space's dimensionality in which both methods are taking place.

While the original technique operates on 2D shapes, the proposed one considers 3D objects. Moreover, in 2D, skeletons have been extracted from the contours, whereas this approach exploits a set of unstructured 3D curve segments for representing the 3D input data. Another variation concerns the estimation of the final correspondences configuration and thus, the overall similarity between two 3D objects.

In contrast to the PSSGM, the cost matrix is not processed by the Hungarian method. Instead of this, a novel matching approach is employed, namely Maximum Weight Cliques (MWC) as introduced in Section 2.4.2.2. Using this approach for the goal of aligning feature points, allows the establishment of partial matchings. The required adaptations towards the MWC, in order to fit the demands of this project, are discussed in the following. Therefore, the query and the target are represented by two graphs G and G' with $K+1$ and $N+1$ nodes, respectively, fulfilling the condition: $K \leq N$.

Unary Potentials These relations are located on the diagonal ($\mathbf{A}_{i,i}$) of the $KN \times KN$-affinity matrix and are populated by the matching costs of each feature pair. However, compared to the Hungarian approach, these values have to express the similarity of the alignment instead of its costs. For this purpose, an ordinary Gaussian function is exploited. As a quick reminder, the actual cost values are obtained by employing the Optimal Subsequence Bijection (OSB) for the task of matching two time series. By means of this method, two features can be matched elastically by skipping possible outliers inside their path sequences. The obligatory *jumpcost* parameter is configured as originally proposed as the sum of the average and the standard deviation (cf. Section 2.4.3.2).

Binary Potentials These quantities form the counterpart to the *unary potentials* and are used to express the consistency between two assignments $r^{\star} = (v,v')$ and $r^{\circ} = (u,u')$. In [ML12], this consistency is defined as spatial distance between the nodes encompassed by the matchings: $v,u \in G$ and $v',u' \in G'$:

$$\mathbf{A}^{(r^{\star},r^{\circ})} = \exp(\frac{(f(v,u) - f(v',u'))^2}{2\theta^2}) \quad , \tag{5.19}$$

where $f(\cdot,\cdot)$ calculates the Euclidean distance and θ adjusts the influence of geometrical deformations on the result. Finally, all values are normalised to a range of $[0,1]$ in order to remain scale invariant. Therefore, the corresponding object height is taken into account.

Mutual Exclusion Constraints As already stated in Section 2.4.2.2, these constraints are crucial for the entire matching process. On the one hand, they are capable of reducing the search space, a fact that improves the computation performance remarkably. On the other hand, they are strengthening the power of the binary potentials. The constraints by itself are implemented as logical geometrical restrictions expressing the relative spatial offset between two feature points: $\mathbf{p}_i, \mathbf{p}_j \in (\mathbb{D}^C)^{(\text{final})}$ and $i \neq j$. Therefore, the object's LCS, introduced in Section 5.3.1, is exploited to support the establishment of these relations in terms of *left/right*, *up/down* and *front/back*. Finally, the features are projected on the axes $\dot{\mathbf{x}}$, $\dot{\mathbf{y}}$ and $\dot{\mathbf{z}}$ of the LCS: $\hat{p}_{\{i,j\}}^{(x)}$, $\hat{p}_{\{i,j\}}^{(y)}$ and $\hat{p}_{\{i,j\}}^{(z)}$. By inspecting the differences of these values, e.g. $|\hat{p}_i^{(x)}| - |\hat{p}_j^{(x)}|$, the relative position of both points can then be easily determined with respect to a certain dimension.

Taking now two assignments $r^* = (v, v')$ and $r^\diamond = (u, u')$, the geometrical relations between its graph nodes ($v, u \in G$ and $v', u' \in G'$) are calculated as described above. Only by monitoring the sign of both differences, the algorithm is able to exclude this configuration from the final result. However, if (v, u) or (v', u') are close to each other so that their distance drops below a certain threshold, the affected dimension is not further considered. In these cases the potential risk exists that the points are changing their relative position caused by measuring errors or inaccuracies.

5.7 Summary

This project of the present thesis introduced a 3D object retrieval system capable of categorising 3D objects represented by 3D curve fragments. For this purpose, a sophisticated scene analysis has been performed utilising a depth device (a Kinect© sensor) able to capture depth images of an arbitrary scene. Afterwards, these depth maps were passed to an automatic detection mechanism as proposed in [Yi+13], leading to a list of scene objects. In order to convert these scene segments to sets of unstructured 3D curves as illustrated in Figure 5.3a, the depth images have been processed unsupervised according to [MYL13]. Therefore, a Canny edge detector is first applied on the pure depth image. The resulting edge fragments are then passed to an edge tracking procedure with the aim to generate 2D curves by linking the edges to disjoint subsets. Subsequently, the detected 2D curves are back-projected into 3D space. This step is realised based on the reconstructed 3D point cloud straightly derived from the same depth image as used before. The 3D correspondences are obtained by collecting those 3D points which belong to the pixels modelling the 2D curve. Once all points are gathered, this set is processed by iteratively applying the RANSAC approach to it.

Moreover, a technique has been proposed combining the actual retrieval process with the capability of accepting user-sketched queries as shown in Figure 5.3b. This aspect especially constitutes a great benefit compared to other methods, e.g. [LFU13; Bo+11] and [Blu+12] since it does not require any expensive or time consuming training phases. Additionally, the highly abstracted representation type, namely the 3D curves, leads to a lower complexity compared to surfaces. In consequence of this, an easy to handle user interface can be provided enabling the construction of search requests based on simple line primitives. Hence, the lines are used to sketch the input queries on the one hand and on the other, they are utilised to approximate the object's geometry. The core idea of this concept had been introduced by authors Ma et al. [MYL13]. In contrast to them, this work considers the 3D curve fragments as a special kind of skeleton in order to employ a matching technique in analogy to [BL08]. As an own contribution, the project introduced: (i) the estimation of a local coordinate system, (ii) the detection of geometry characterising feature points which significantly contribute to the perceptual appearance of an object, (iii) a RST invariant path descriptor and (iv) an adaptive, unsupervised 3D point ordering scheme. Additionally, an extended version of the Dijkstra algorithm has been proposed with the intent to tackle ambiguities caused by symmetry and measurement errors during the shortest path computation.

Summarised, the method led to excellent results which even outperformed the original work presented in [MYL13]. These numbers have been obtained by a thorough evaluation presented in Section 8.3. The results confirm the proposed method a robust behaviour even in presence of obstacles like occlusions, distortions, measurement inaccuracies and outliers. However, only by incorporating the matching approach based on maximum weight cliques, the recognition precision has been increased to a superb level of accuracy. Nevertheless, the current technique is restricted to a specific amount of objects whose geometry is neither too complex nor entirely opaque. These limitations are further stressed due to the fact that the scene acquisition follows a *fixed view* setup which suffers from the inherent problem that only the object's front can be captured. Strictly speaking, the algorithm is not able to obtain any information about the object's back. For the time being, this data is lost and will not be considered in subsequent tasks. Nevertheless, the future work focuses on the inclusion of a registration-driven strategy capable of merging multiple sets of 3D curves displaying the same object from different perspectives. Moreover, the object representation will be extended to continuously curved 3D fragments in order to enable further objects geometries. Consequently, the database can be extended with the intent to address further application domains.

Chapter 6

Skeleton-based Abdominal Aorta Registration

> "Vascular disease is a serious disease of the arteries and veins that blocks
> [blood] circulation anywhere in the body. Vascular disease is serious and can
> lead to disability, amputation, organ damage and even death."[1]

Vascular diseases are one of the most challenging health problems (even) in developed countries. According to statistics acquired by the Centers for Disease Control and Prevention (CDC), aortic aneurysms (AAs) have been responsible for around 10000 deaths only in the United States in 2009[2]. In the United Kingdom, the number of deaths is not as high, but is still at around 7000 as indicated by the British Heart Foundation (2011)[3]. In Germany, statistics of the Federal Statistical Agency (Statistisches Bundesamt) revealed about 3500 deaths caused by AAs in the year 2012[4]. Hence, vascular segmentation as well as registration techniques are the topics of past and are still highly important on-going research activities.

Definition 6.0.1. *An aortic aneurysm describes an artery blockage disease that occurs inside the aorta. Figure 6.1 shows the most common types of this disease, namely (i) the thoracic AA located behind the heart and (ii) the abdominal AA close to the vessels supplying blood to the kidneys. Among these two types, the abdominal AAs appear more frequently than the thoracic ones. As illustrated, aneurysms are balloon-like dilatations inside the aorta. They increase the aorta diameter permanently by at least 50% of their expected size [Abr+09] (approximately 2 cm). Without treating them appropriately, the thickness of the aorta wall becomes weak and can rupture due to the normal blood pressure. The (impending) rupture of the aorta wall leads to pain, haemorrhage and to death.*

---

[1]http://vasculardisease.org, [online: 19th August 2015]
[2]http://www.cdc.gov, [online: 19th August 2015]
[3]https://www.bhf.org.uk, [online: 19th August 2015]
[4]http://www.gbe-bund.de, [online: 19th August 2015]

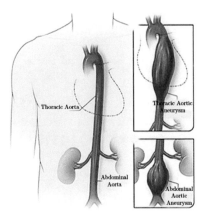

Figure 6.1: The figure shows the two most common types of aortic aneurysms[5]. On the one hand, this is the *thoracic* AA located behind the heart and on the other hand the *abdominal* AA close to the vessels supplying blood to the kidneys.

The following content targets the problem of abdominal AA for the purpose of representing a registration technique using the Path Similarity Skeleton Graph Matching (Section 3.2) and a modification of it that incorporates Maximum Weight Cliques as introduced in Section 2.4.2.2 (cf. [Fei+14d; Cza+14]). Firstly, the *aorta lumen* is segmented as described in Section 6.2.1. Secondly, the centre line is extracted from this structure (Section 6.3) which is then finally used in Section 6.4.1 and Section 6.4 for realising the actual *graph matching* technique to register pre- with post-operative CTA series.

In particular, this approach is developed in the light of supporting the physician during his after-care of abdominal aortic aneurysms by visualising correspondences between pre- and post-operative vascular structures. Compared to the projects before, this problem is more sophisticated and differs in multiple aspects: (i) Size and resolution of the voxelised data as well as (ii) the number of slices might be different between the scans, (iii) the region of interest might be shrunk or enlarged, (iv) branches might disappear or occur caused by the resolution variability affecting the topology of the centre line. On the one hand, all of these issues increase the complexity of the registration problem, but on the other, their existence emphasises the power and universality of the method presented in this chapter.

The actual evaluation has been performed on a database of real-world patients taking

[5]http://www.cdc.gov/dhdsp/data_statistics/fact_sheets/fs_aortic_aneurysm.htm, [online: 19th August 2015]

into account two contrast-enhanced CTA series for each person. All configurations have been verified by an expert qualified to assess the anatomical correctness of this outcome. Furthermore, promising results confirm the system an excellent performance and encourage further activities on this topic at the same time.

6.1 Fundamental Concepts

6.1.1 Introduction to the level set Method

"The level set method is just plain easy to understand: there is a surface, it intersects a plane, that gives us a contour and that's it."[6]

The concept of level set methods was introduced by Osher and Sethian [OS88] in 1988 having a huge impact on many areas of computational science. The level set approach has to be considered as a highly accurate and robust movement tracking mechanism of interfaces even under complex motion. In context of this work, it is used for 3D object segmentation as well as for the purpose of 3D skeletonisation. However, before these topics are subject of a detailed discussion, the principle of the native *curve evolution* shall be explained. Compared to other techniques in this area, the implicit formulation of an interface as a constant set in higher dimensional space (the *zero level set*) offers some remarkable advantages:

"(1) [...] topological changes of the contours can easily be handled; (2) the concept and numerical implementation can be adapted to solve any dimensional problems; (3) the areas inside and outside an active contour are distinguishable." [Zha+08]

The transformation of an explicitly parametrised curve C into an implicit zero level set formulation can be expressed as follows:

$$C = \{\mathbf{x} \mid \phi(\mathbf{x}) = 0\}, \tag{6.1}$$

where the interface ϕ is defined over a domain Ω separated by the curve into two regions (Ω^+, Ω^-) with zero elements on the border (zero level set). Moreover, splitting and merging operations are natively handled during the evolution process, and thus, topological changes are registered automatically. The actual evolution of the interface is realised in higher dimensional space, e.g. a curve is defined as the zero level set in 2D. Furthermore, the modelling of such an evolution process requires a time-dependent parameter t in addition

---

[6]http://step.polymtl.ca/~rv101/levelset/, [online: 19th August 2015]

to the spatial component **x** indicating the single steps during this evolution. This temporal correlation can be expressed by reformulating Equation (6.1) with the intention of adding a second parameter:

$$C = \{\mathbf{x} \mid \phi(x(t), t) = 0\} \quad , \tag{6.2}$$

$x(t)$ is a function over the time returning the position of an arbitrary point on the curve at any time t. Moreover, the actual level set function ϕ can freely be chosen as long as it determines the position of the evolving curve C. In other words, the zero level set has to coincide with the set of points belonging to this curve: $\mathbf{x}_i \in C$.

The actual evolution of such an interface requires two inputs, namely (i) an initial zero level set ($\phi(x(t), t = 0) = 0$) and (ii) a corresponding motion equation ($\partial\phi/\partial t$). The latter is obtained as derivative of ϕ using the chain rule in respect of the time:

$$\frac{\phi(x(t), t)}{\partial t} = 0 \tag{6.3}$$

$$\frac{\partial\phi}{\partial t} + \frac{\partial\phi}{\partial x(t)} \cdot \frac{\partial x(t)}{\partial t} = 0 \tag{6.4}$$

$$\phi_t + \nabla\phi(x(t), t) \cdot x'(t) = 0 \quad . \tag{6.5}$$

Knowing that $\partial\phi/\partial x = \nabla\phi$ and that $x'(t)$ determines the speed, Equation (6.5) can be reformulated to the fundamental *level set function (LSF)*, where v depicts an arbitrary velocity field steering the speed of the evolution process:

$$\phi_t + v\nabla\phi = 0 \quad . \tag{6.6}$$

Furthermore, (6.6) forms a first-order partial differential equation (PDE) and belongs to the more general set of Hamilton-Jacobi equations. In a discretised space, e.g. the Cartesian grid, the finite difference scheme can be employed to solve this PDE numerically. Typically, the evolution process is performed iteratively until a so-called *steady state solution* is reached. A state where no further changes appear or, in other words, where further changes are below a certain threshold.

The LSF (ϕ) is most frequently implemented as a so-called signed distance function (SDF) like exemplarily defined in Equation (6.7) satisfying the property $|\nabla\phi| = 1$:

$$\phi(\mathbf{x}) = \begin{cases} D(\mathbf{x}), & \forall \mathbf{x} \in \Omega^+ \\ -D(\mathbf{x}), & \forall \mathbf{x} \in \Omega^- \end{cases} \quad , \tag{6.7}$$

where Ω denotes the domain in which the interface is defined, $D(\cdot)$ is the distance transform (DT) calculating the distance of each grid point to the curve (cf. Figure 6.2). In consequence, all differentiable points x_i on the interface C fulfil the condition: $\nabla\phi = \mathbf{n}$. Later in this section, the benefit of this property is going to be discussed in more detail. For the time being, it is worth mentioning that keeping the LSF close to $|\nabla\phi| = 1$ results in a higher numerical stability as well as evolution accuracy. Strictly speaking, the LSF should be smooth in its structure. Unfortunately, this task is not trivial due to external motion constraints (e.g. curvature) capable of corrupting the shape of the SDF. Thus, complex techniques have to be applied in order to recover it. This process is better known as *reinitialisation*.

Without using reinitialisation the accuracy drops and numerical errors are accumulated over time. Since the evolution is a continuous process, the reinitialisation has to be invoked periodically during the evolution. A bunch of different methods exist (cf. [HMS08]), able to take care of the SDF's shape. Besides the approach of reinitialisation as post-processing step, other techniques directly apply regularisation terms to the PDE [Li+05; Li+10].

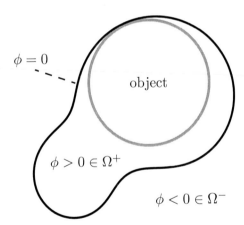

Figure 6.2: The figure shows a possible curve evolution at a certain time. Values inside the curve are assigned to Ω^+, while the values outside are members of Ω^-. All points on the curve have zero distance and thus, they are known as the zero level set (cf. [Zha+08]).

6.1.1.1 Normal-driven Curve Evolution

The normal-driven curve evolution is a special instance of the more generic level set function shown in Equation (6.6). In particular, this kind of curve evolution is frequently used in the field of image processing, e.g. for the task of segmentation. Instead of using an external velocity field, the curve evolution is performed based on the normals of the interface. In order to establish this behaviour, the following pre-condition can be assumed: $v = \psi \mathbf{n} = \psi \nabla \phi / |\nabla \phi| \, (= x'(t) \cdot \mathbf{n})$ with ψ being the speed of the curve flow in normal direction to the curve[7]. Thus, by substituting v in (6.6), the equation changes to:

$$\phi_t + \psi \frac{\nabla \phi}{|\nabla \phi|} \nabla \phi = 0 \qquad (6.8)$$

$$\phi_t + \psi |\nabla \phi| = 0 \quad , \qquad (6.9)$$

which is known as the standard level set technique having a motion in normal direction. Recalling the properties of an SDF, this equation can be further relaxed by taking $|\nabla \phi| = 1$ into account.

$$\phi_t + \psi = 0 \quad . \qquad (6.10)$$

Please notice that the speed can be defined in multiple ways, e.g. it might be implemented based on certain image characteristics (intensity values) and/or by utilising the curvature of the interface.

6.1.1.2 Finite difference scheme

The finite difference scheme constitutes a popular instrument to solve a PDE numerically[8] as already mentioned above. Thus, it is frequently used as implementation with the intent to perform the actual curve evolution. In the following, the first-order *upwind scheme* shall briefly be discussed. This scheme exploits finite differences for the approximation of both the temporal and the spatial derivative based on the corresponding LSF. Please remember that the focus is still on the model equation (cf. [Pat10]):

$$\frac{\partial \phi}{\partial t} + v \frac{\partial \phi}{\partial x} = 0 \quad . \qquad (6.11)$$

---

[7]http://www.cs.umd.edu/~djacobs/CMSC828seg/LevelSets.pdf, [online: 19th August 2015]
[8]http://www.ehu.eus/aitor/irakas/fin/apuntes/pde.pdf, [online: 19th August 2015]

Temporal Derivative   In order to compute the temporal derivative, the *forward difference* is employed. With access to this difference scheme, Equation (6.6) can be transformed as follows:

$$\frac{\phi^{k+1} - \phi^k}{\Delta t} + v^k \nabla \phi^k = 0 \quad , \tag{6.12}$$

having in mind that $\phi^k = \phi(x(t^k), t^k)$ and $\Delta t = t^{k+1} - t^k$, the equation can be rearranged to:

$$\phi^{k+1} = \phi^k - \Delta t \, (v^k \, \nabla \phi^k) \quad . \tag{6.13}$$

Closely inspecting Equation (6.12) and (6.13), one can see that the concept behind this scheme is quite straightforward and easy to apply.

Spatial Derivative   For the purpose of deriving the spatial component, a combination of two finite difference schemes is recruited to approximate the gradient at each position. The meaning is about the *forward* and the *backward difference scheme* which are frequently used in literature. With the intention of simplification, the following explanations are reduced to only one dimension with ϕ_x being the derivation of the spatial component at t^k.

$$\phi^{k+1} = \phi^k - \Delta t \, (v^k \, \phi_x^k) \quad . \tag{6.14}$$

Please notice that in a multi-dimensional space each dimension is evaluated the same way. The overall idea is to control the gradient direction relying on v_i^k by a combination of the *forward* and the *backward difference* scheme:

$$\phi_x = \begin{cases} \frac{\phi_{x+1}^k - \phi_x^k}{\Delta x} = \phi_x^+, & v^k < 0 \\ \frac{\phi_x^k - \phi_{x-1}^k}{\Delta x} = \phi_x^-, & v^k > 0 \end{cases} \quad . \tag{6.15}$$

Another notation which is frequently used can be written as follows: $\phi^{k+1} = \phi^k - \Delta t \left(\min(v^k, 0) \phi_x^+ + \max(v^k, 0) \phi_x^- \right)$, where $\max(\cdot, \cdot)$ and $\min(\cdot, \cdot)$ are functions returning respectively the maximum and the minimum value in respect of the given input parameters.

6.1.1.3   Reinitialisation

As already introduced above, the signed distance function constitutes an appropriate means for the goal of implementing a curve evolution process. However, besides many attractive properties arguing for its use, the SDF's shape requires regular updates during the actual evolution procedure. These update cycles are the result of the dynamic model engendering numerical errors which are affecting the look of the function with the consequence that $\|\nabla\phi\| = 1$ is no longer valid. Strictly speaking, the gradients, or in other words, the slope of the function becomes too steep or flat and the evolution is running the risk of hitting on problems regarding its numerical stability as well as its accuracy (cf. [Joh11]). This smoothing or recovery process is known as *reinitialisation*.

> "Everytime we reinitialize, the interface moves, so reinitialization should only
> be done when necessary, as too frequent reinitializations will cause the interface
> movement to have too big of an impact on the final results, but too few will
> cause the level set function to lose its good properties, with loss of accuracy as a
> result." [Joh11]

The easiest way to implement such a reinitialisation is the brute force approach that calculates all distances to the interface by considering all points of the underlying domain. Like other brute force implementations, such a reinitialisation is highly time consuming. Instead of this, a more practical and more sophisticated technique is proposed by Sussman et al. in [SSO94]:

$$\phi_t + \psi(\tilde{\phi})(|\nabla\phi| - 1) = 0 \quad , \tag{6.16}$$

with $\psi(\tilde{\phi})$ being the smoothed distance function which is defined as:

$$\psi(\tilde{\phi}) = \frac{\tilde{\phi}}{\sqrt{\tilde{\phi}^2 + \epsilon^2}} \quad . \tag{6.17}$$

Please notice that the parameter ϵ is used to steer the amount of smoothing and has to be selected carefully. In theory, Equation (6.16) has to be solved until a steady state is reached. Practically, the equation is only applied to those regions close to the interface based on the assumption that further evolution steps are restricted to this area (cf. [HMS08]).

Fast Marching Method   The FMM [Set95; Set99a] is another technique to enforce the smoothness of an SDF. In literature, this method is widely used for implementing level set-driven applications like the segmentation (Section 6.2.1) and the skeletonisation (Section 6.3) used in this project.

> "FMMs are numerical schemes for computing solutions to the nonlinear
> Eikonal equation and related static Hamilton-Jacobi equations. Based on entropy-

satisfying upwind schemes and fast sorting techniques, they yield consistent, accurate, and highly efficient algorithms" [Set99a].

The core idea behind the FMM is solving the *Eikonal* equation with the aim to determine the gradient lengths of the time crossing map $\|\nabla T\|$. Therefore, the actual evolution of the interface is formulated as follows (known as Eikonal equation, cf. [Joh11]):

$$v\|\nabla T\| = 1 \quad , \tag{6.18}$$

where the distance of a point to the interface is interpreted as the time of travelling. Considering an arbitrary location â of the underlying domain, the function $T(\cdot)$ returns the crossing or rather the arrival time of the interface with/at â as demonstrated in Figure 6.3. In addition to this, $v(\cdot)$ indicates the motion speed of the interface in normal direction.

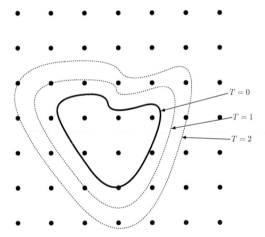

Figure 6.3: The figure demonstrates the curve evolution to certain time steps. Moreover, it illustrates the notion behind the concept of *time crossing* between a curve and the grid points of a domain (cf. [Bær01]).

By setting the speed to $v = 1$, the time crossing map can easily be calculated by solving $\|\nabla T\| = 1/v \equiv \|\nabla T\| = 1$ with the result of becoming a signed distance function. The actual gradient length is then approximated according to the upwind scheme (here 1D):

$$\|\nabla T\|^2 = \max(T_x^-,0)^2 + \min(T_x^+,0)^2 \quad , \tag{6.19}$$

where $T_x^{\{+,-\}}$ respectively determines the forward and backward differences at a certain position x (cf. Equation (6.15)). By passing these differences to the corresponding minimum/maximum functions, the larger upwind solution is automatically selected. In a multi-dimensional case, the remaining dimensions are processed in the same way. Finally, the desired solution is obtained by solving the quadratic term of this equation. If two different results are retrieved, the larger one has to be preferred in order to keep the monotony criteria.

As trivial as this technique might appear, it suffers from measurement inaccuracies caused by the first-order derivatives. Moreover, the error increases with the degree of curvature. Being aware of this problem, further approaches have been proposed. In [Set99b], Sethian developed a more accurate version by replacing the first-order derivatives with second-order ones. Apart from this, the authors in [HF07a] proposed a multistencils approach that even outperforms this method (cf. Section 2.2.6).

6.1.2   Introduction to the k-Means Clustering Approach

The k-means clustering algorithm is a popular instrument in the field of machine learning and data mining. Originally proposed by James MacQueen in 1967 [Mac67], plenty of variations have been derived from this technique up to now. For the sake of completeness, some of them are mentioned in the following: The k-medians clustering, the fuzzy c-means clustering and the Gaussian mixture models. Most of these variations try to tackle the initialisation problem as also discussed later in this chapter. Even though the fundamental problem is NP-hard, the core idea of the algorithm is trivial. Given an arbitrary data set of observations $\mathbb{D} = \{\bar{x}_1, \bar{x}_2, \dots, \bar{x}_N\}$, the algorithm tries to separate $\mathbb{D}$ into $k \leq N$ partitions, so that the sum of squared distances to each cluster $\hat{q}_i$ is minimal. This optimisation task is then be formulated as:

$$\underset{Q}{\arg\min} \sum_{i=1}^{k} \sum_{\bar{x}_j \in Q_i} \|\bar{x}_j - \hat{q}_i\|^2 \quad , \tag{6.20}$$

with Q being the set of possible cluster configurations. Moreover, the method belongs to the class of unsupervised mechanisms since it does not require any labels on the input data. The most famous implementation was proposed in [Llo82] determining the optimal solution in an iterative manner.

However, there are some potential problems which are challenging to solve. First, the clustering result strongly relies on the initial configuration and thus, it cannot be guaranteed (without incorporating a priori knowledge) to find the best solution. Second, the result is correlating with the expected number of clusters which has to be determined beforehand. Although further problems are noted, one modified version is going to be introduced later in this section addressing most of these known issues.

6.2 Segmentation of the Abdominal Aorta, the Kidneys and the L4 Segment of the Lumbar Spine

"Image segmentation methods using active contours are usually based on minimising functionals which are so defined that curves close to the target boundaries have small values." [Zha+08]

Please remember that this chapter is aiming at the registration of vessel structures coming from the abdominal aorta for the purpose of supporting medical diagnosis. Therefore, the body's interior is first acquired using a Computed Tomography scanner. This output is then passed to the segmentation stage for the extraction of the pure vessel tree. Finally, before the actual matching procedure can start, this data is skeletonised. While the skeletonisation is subject of Section 6.3 and the registration approach of Section 6.4, this content is devoted to the segmentation introduced by the authors Zhang et al.

6.2.1 Segmentation of the Abdominal Aorta Using Active Contours

The segmentation process extracts the abdominal aorta by employing an active contour-driven level set approach, proposed by the authors Zhang et al. [Zha+08]. The actual implementation of this method exploits an *energy function* $\mathcal{E}(\phi)$ having a smaller value the better the curve (ϕ) characterises the target boundary of a specific object. The actual curve is embedded implicitly as the zero level set as discussed in Section 6.1.1. In order to minimise the energy $\mathcal{E}$, the steady state solution of the gradient flow equation is searched. Therefore, a PDE is constructed as Gâteaux derivative of the functional $\mathcal{E}$ by $\frac{\partial \phi}{\partial t} = -\frac{\partial \mathcal{E}}{\partial \phi}$. The actual energy function $\mathcal{E}$ includes the properties of the boundary and the image region:

$$\mathcal{E}(\phi) = -\eta \int_\Omega (\mathbf{I} - \mu^{(\text{scope})}) \, H(\phi) \, d\Omega + \tau \int_\Omega g |\nabla H(\phi)| \, d\Omega \quad . \tag{6.21}$$

The region characteristics are accommodated inside the first term. The predefined parameter $\mu^{(\text{scope})}$ restricts the lower intensity level of the target object inside the image $\mathbf{I}$ to enforce the curve to enclose those regions with intensity values above $\mu^{(\text{scope})}$ during the evolution. In contrast to this, the second term operates on the object's boundary which is obtained by the image gradients accessible by the function $g(|\nabla I|)$ which is typically defined as:

$$g = \exp(-ag|\nabla I|^2) \quad \text{or} \quad g = \frac{1}{1 - a|\nabla I|^2} \quad , \tag{6.22}$$

with a being a constant controlling the slope. Furthermore, the second term of Equation (6.21) is known as *geodesic active contour functional* and has already been introduced by Caselles et al. in [CKS97]. It encourages the curve to attach itself to the areas with high image gradients (cf. [Zha+08]). The Heaviside or step function depicted by H (added to each term) prompts the evolution to operate only on the image data that is currently covered by the curve (in a certain evolution step).

$$H(\phi) = \begin{cases} 0, & \phi < 0 \\ 1, & \phi \geq 0 \end{cases} . \tag{6.23}$$

Finally, η and τ are two predefined weights balancing the region to the boundary term and vice versa. The benefit of this hybrid method lies in its robustness and accuracy. While the boundary information supports a precise location of the target object, the region data prevents boundary leakage problems.

The actual PDE is finally obtained by deriving Equation (6.21) with respect to the embedded level set function (ϕ). It is worth mentioning that $\delta(\phi)$, known as Dirac function, is used as derivative of the heaviside function and thus, $\delta = H'$ whereas div($\cdot$) denotes the divergence operator:

$$\phi_t = \delta(\phi)\left(\eta\,(\mathbf{I} - \mu^{(\text{scope})}) + \tau\,\text{div}\Big(g\frac{\nabla\phi}{|\nabla\phi|}\Big)\right) \quad . \tag{6.24}$$

Knowing that $\delta(\phi)$ can also be replaced by $|\nabla\phi|$ and recalling that the signed distance function has the property $|\nabla\phi| = 1$, Equation (6.24) can be further relaxed to:

$$\phi_t = \eta\,(\mathbf{I} - \mu^{(\text{scope})}) + \tau\,\text{div}(g\nabla\phi) \quad . \tag{6.25}$$

Finally, the foregoing PDE is updated according to the explanations given in Section 6.1 coupled with an appropriate reinitialisation scheme taking care of the SDF's shape ($|\nabla\phi| = 1$). Figure 6.4 illustrates one of the segmentation results showing a fully extracted abdominal aorta obtained by the approach state above.

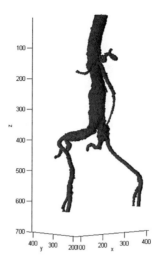

Figure 6.4: Segmentation result of the abdominal aorta using active contours.

6.2.2   Segmentation of the Kidneys and L4 of the Lumbar Spine

A closer look at Figure 6.4 reveals that the aorta structure does not provide highly distinctive characteristics and this, in turn, is problematic in context of registering these vessels. Consequently, the use of Maximum Weight Cliques (MWC) (as introduced in Section 2.4.2.2) offers a highly appropriate instrument to improve the robustness of this process. The benefit of exploiting this method is its capability of taking into account further information in form of mutual exclusion (mutex) to enrich the actual matching procedure. In order to establish these constraints the decision was made to utilise three further organs of the human body, namely the two kidneys and L4 of the lumbar spine. Subsequently, these organs are considered as reference points for both the generation of an appropriate feature set as well as the introduction of a local coordinate system. The use of these organs is reasonable since they provide a stable orientation inside the human body.

   This section briefly explains the chosen segmentation approach which is better known as *Kernelised Fuzzy C-Means* clustering. The technique is applied on the images of CTA series as proposed in [KP07] with the goal to retrieve the location of the desired organs.

6.2.2.1 Fuzzy C-Means Clustering

Before the actual segmentation topic concerning the *Kernelised Fuzzy C-Means* clustering can be discussed, this section is dedicated to the original *Fuzzy C-Means* algorithm. The latter, in turn, constitutes an extension of the k-means method as discussed in Section 6.1.2. As the name already suggests, the *Fuzzy C-Means* technique enhances the k-means approach by the component of *fuzzy logic*.

Fuzzy Logic   The fuzzy logic was first introduced by Lofti A. Zadeh in 1965. The fuzzy set theory must be considered as a *many-valued logic* which allows the modelling of systems having more than only two states. The fuzzy logic is an important instrument to describe *unsharp* conditions which are not representable based on the binary set {*true, false*}. In contrast, the fuzzy logic provides several degrees of "truth" which are mapped into a range varying between *true* and *false*.

The fuzzy part counteracts the hard or the crisp assignment problem occurring in context of the native k-means algorithm. Strictly speaking, it alleviates the restriction of assigning each observation to only one cluster. Thus, a single observation can be the member of multiple partitions $Q_{1,...,k}$ at a time. This is realised by the *membership matrix* describing the degree of affiliation to a certain cluster. Let $\mathbb{D} = \{x_1, x_2, \ldots, x_N\}$ be a set of observations and $\hat{q}_i \in Q$ be the cluster representatives. By augmenting the objective function as depicted in (6.20) with the membership coefficients or partition elements $\psi_{i,j}$, the minimisation of the new cost function is expressed as:

$$\arg\min \sum_{i=1}^{k} \sum_{j=1}^{N} \psi_{i,j}^{h} \|x_j - \hat{q}_i\|^2 \quad . \tag{6.26}$$

In this context, one of the most interesting parameters is given by h determining the fuzzyfication level in a range of $]1, \infty[$, whereas the membership coefficients are in $[0, 1]$. Moreover, the procedure is not allowed to produce empty clusters and thus, the following condition has to be satisfied at any time: $(\sum_{j=1}^{N} \psi_{i,j} > 0)$. The actual minimisation is then realised by the well-known *Lagrangian Multipliers* that ensure the adherence of the conditions stated above. Therefore, Equation (6.26) has to be derived partially resulting in two different formulas which can be used in an iterative optimisation scheme to find the optimal solution:

$$\psi_{i,j}^{t+1} = \Big(\sum_{i'=1}^{k} \Big(\frac{\|\mathbf{x}_j - \hat{\mathbf{q}}_i^t\|}{\|\mathbf{x}_j - \hat{\mathbf{q}}_{i'}^{t-1}\|} \Big)^{\frac{2}{h}} \Big)^{-1} \quad , \tag{6.27}$$

$$\hat{\mathbf{q}}_i^t = \frac{\sum_{j=1}^{N} \psi_{i,j}^{t,h} \, \mathbf{x}_j}{\sum_{j=1}^{N} \psi_{i,j}^{t,h}} \quad . \tag{6.28}$$

Having these two formulas, the optimisation additionally requires an initial set of partition coefficients $\psi_{i,j}^{(0)}$ in order to perform iteratively until the maximum movement towards all cluster centres is below a user defined threshold.

6.2.2.2 Kernelised Fuzzy c-Means Clustering

After this short introduction into the topic of fuzzy c-means clustering, the following content is dedicated to the actual segmentation method finally employed for the purpose of organ segmentation, namely the *kernelised fuzzy c-means clustering*. The approach has been first presented in [KP07] by Kawa and Pietka and extends the previously explained *fuzzy c-means Clustering* (cf. Section 6.2.2.1). Thus, it can be viewed as an extension of the extension regarding the k-means method introduced in Section 6.1.2. In its core, the technique maps the concept of fuzzy clustering into a *kernel space*.

Kernel Space   This space is well-known in the field of data analysis and highly suitable for machine learning. It is frequently employed in many scenarios as [CNT13] and [LZL04] in order to mention some of them. Operating in this space usually provides attractive properties, e.g. in terms of separability. The most popular example in this context is the *Support Vector Machine* [TK08]. Hence, the notion behind using a kernel space-driven approach lies in the fact that data is often more easily to separate in higher dimensional space than inside the embedding one. Moreover, this way non-linear separation problems are transformed into linear ones only by increasing the dimensionality of embedding space.

The mapping is typically realised by exploiting the *kernel trick*. In more detail, the kernel trick represents an easy to implement mathematical relaxation which is commonly used for this purpose. Moreover, it provides attractive computational properties in terms of memory and power consumption. The *kernel function* ($\hat{k} : \mathcal{D} \times \mathcal{D} \to \mathbb{R}$), or only *kernel*, can be considered as the inner product of two data vectors $\mathbf{x}$ and $\mathbf{x}' \in \mathcal{K}$ mapped into a higher dimensional space $\mathcal{H}$. The space $\mathcal{H}$ is also known as *Hilbert Space* and generalises the concept of the Euclidean vector space by the capability to encompass any number of dimensions. "A Hilbert space is [an abstract] vector space with an inner product [...] such that [it constitutes] a complete metric space"[9]. Thus, measurements in terms of length and angle are valid operations. Assuming that $\hat{k}(\cdot,\cdot)$ is positive definite, a function $\varrho(\cdot)$ exists enabling the following kernel definition:

$$\hat{k}(\mathbf{x},\mathbf{x}') = \varrho(\mathbf{x})^\top \varrho(\mathbf{x}') \quad , \tag{6.29}$$

where $\varrho : \mathbb{R}^n \to \mathcal{H}$ maps $\mathbf{x}$ and $\mathbf{x}'$ into a higher dimensional space $\mathcal{H}$. The mapping $\varrho(\cdot)$ does not necessarily have to be known as long as $\mathcal{H}$ meets the requirements of being an inner product space. It is worth mentioning that the kernel $\hat{k}(\cdot,\cdot)$ is symmetric and that it satisfies the following condition according to its property of being positive definite:

$$\sum_{i=0}^{N-1} \sum_{j=0}^{N-1} \tau_i \tau_j \hat{k}(\mathbf{x}_i,\mathbf{x}_j) \geq 0 \quad , \tag{6.30}$$

with N indicating the number of observations and τ_i, τ_j representing N real-valued coefficients. Bear in mind that this trick constitutes the basis of classifying non-linear separation problems based on a linear classifier.

Kernelised Fuzzy Clustering   In [KP07] the kernel trick is used to represent the clusters implicitly inside a kernel space with the intention of increasing both the robustness and the accuracy of the clustering. In the following, those modifications are introduced which have been proposed by Kawa et al. to incorporate the kernel trick into the fuzzy clustering approach. To accomplish this goal, two requirements have to be fulfilled: First, an appropriate mapping function $\varrho(\cdot)$ exists and second, the kernel $\hat{k}$ is well defined. According to the previously obtained knowledge and by assuming the existence of a valid mapping into $\mathcal{H}$, the objective function of (6.26) can be rewritten as:

$$\operatorname{argmin} \sum_{i=1}^{k} \sum_{j=1}^{N} \psi_{i,j}'^h \|\varrho(\mathbf{x}_j) - \hat{\mathbf{q}}_i'\|^2 \quad . \tag{6.31}$$

As in the previous section, this optimisation problem can be solved iteratively. Therefore, all membership coefficients have to be updated first in order to refine the cluster centres inside the kernel space:

---

[9]Example of a Hilbert space are the real numbers with the vector dot product. http://mathworld.wolfram.com/HilbertSpace.html, [online: 19th August 2015]

$$\psi_{i,j}^{\prime t+1} = \frac{1}{\sum_{i'=1}^{k} \left(\frac{\|\hat{\mathbf{q}}_i^{\prime t} - \varrho(\mathbf{x}_j)\|}{\|\hat{\mathbf{q}}_{i'}^{\prime t} - \varrho(\mathbf{x}_j)\|} \right)^{\frac{2}{h-1}}} \quad , \tag{6.32}$$

and

$$\hat{\mathbf{q}}_i^{\prime t} = \frac{\sum_{j=1}^{N} \psi_{i,j}^{\prime t,h} \varrho(\mathbf{x}_j)}{\sum_{j=1}^{N} \psi_{i,j}^{\prime t,h}} \quad . \tag{6.33}$$

Finally, the kernel trick has to be applied to the formulas derived above. Therefore, $\hat{\mathbf{q}}'$ is substituted in (6.32) by its time-correlating counterpart given in Equation (6.33). Finally, the Euclidean norm has to be transformed into the inner product $\|\cdot\|^2 \equiv \langle \cdot, \cdot \rangle$ leading to the formula below:

$$\psi_{i,j}^{\prime\prime t+1} = \frac{\langle \varrho(\mathbf{x}_j) - \hat{\tau}_i \sum_{j'=1}^{N} \psi_{i,j'}^{\prime t,h} \varrho(\mathbf{x}_{j'}), \varrho(\mathbf{x}_j) - \hat{\tau}_i \sum_{j'=1}^{N} \psi_{i,j'}^{\prime t,h} \varrho(\mathbf{x}_{j'}) \rangle^{\frac{-1}{h-1}}}{\sum_{i'=1}^{k} \langle \varrho(\mathbf{x}_j) - \hat{\tau}_{i'} \sum_{j'=1}^{N} \psi_{i',j'}^{\prime t,h} \varrho(\mathbf{x}_{j'}), \varrho(\mathbf{x}_j) - \hat{\tau}_{i'} \sum_{j'=1}^{N} \psi_{i',j'}^{\prime t,h} \varrho(\mathbf{x}_{j'}) \rangle^{\frac{-1}{h-1}}} \quad , \tag{6.34}$$

with $\hat{\tau}_i = (\sum_{j=1}^{N} \psi_{i,j}^{h})^{-1}$. Here the notation has been borrowed from [KP07] to increase the readability of the formula. Being aware that the inner product can be replaced by the kernel function $(\hat{k}(\cdot, \cdot) = \langle \cdot, \cdot \rangle)$ and by taking into account the property of (6.30), Equation (6.34) can be further transformed to:

$$\psi_{i,j}^{\prime\prime t+1} = \frac{\left(\hat{\tau}_i^2 \sum_{j'=1}^{N} \sum_{r=1}^{N} (\psi_{i,j'}^{\prime} \psi_{i,r}^{\prime})^h \hat{k}_{j',r} - 2\hat{\tau}_i \sum_{j'=1}^{N} \psi_{i,j'}^{h} \hat{k}_{j',j} + \hat{k}_{j,j} \right)^{\frac{-1}{h-1}}}{\sum_{i'=1}^{k} \left(\hat{\tau}_{i'}^2 \sum_{j'=1}^{N} \sum_{r=1}^{N} (\psi_{i',j'}^{\prime} \psi_{i',r}^{\prime})^h \hat{k}_{j',r} - 2\hat{\tau}_{i'} \sum_{j'=1}^{N} \psi_{i',j'}^{h} \hat{k}_{j',j} + \hat{k}_{j,j} \right)^{\frac{-1}{h-1}}} \quad , \tag{6.35}$$

with $\hat{k}_{i,j}$ being the abbreviated notation of $\hat{k}(\mathbf{x}_i, \mathbf{x}_j)$. It is obvious that the method requires a kernel function as input responsible for the generation of the *kernel matrix* $[\hat{k}_{ij}]_{N \times N}$. Moreover, the parameter h (the fuzzification level) and $\hat{\tau}$ have to be known. Given a set of coefficients $(\psi_{i,j}^{\prime\prime(0)})$ initialising the membership matrix, the approach performs iteratively until the maximum distance between the coefficients of two successive time steps is below a pre-defined threshold.

6.3 Skeleton Extraction from the Aorta's Vascular Structure

Recalling that the input data has been processed by multiple stages up to now with the result of two segmented kidneys, the L4 segment of the lumbar spine as well as the abdominal aorta, this section covers the topic of extracting the centre line of the vessel's structure, namely the skeletonisation. Therefore an approach is selected which perfectly connects to the segmentation method introduced in Section 6.2.1 since both techniques are sharing the use of the Fast Marching Method (FMM). While the FMM is employed to generate a distance map in context of the skeletonisation, Section 6.2.1 exploits it for the purpose of reinitialization. The actual extraction of the skeletal structure is worthwhile due to attractive properties especially in context of the object representation (cf. Section 2.4.1). The chosen skeletonisation process was first introduced by the authors Robert Van Uitert and Ingmar Bitter in [VUB07]. Their technique focuses on the generation of skeletons based on subvoxel precision. The benefit of such an accuracy level is illustrated in Figure 6.5, where the higher precision scale is shown in comparison to the one limited to the grid resolution. It is obvious that the latter, drawn as solid line, clearly suffers from discretisation artefacts (cf. Section 7.2). Another attractive property of this technique is its fully autonomous operation principle that does not require any user inputs, e.g. the number of branches or specific object parameters. Please notice that this automatism is highly convenient since the identification of an appropriate number of end points constitutes a sophisticated task with a huge impact on the result's quality. The computational demand for the entire algorithm is estimated with $O(h * n * \log n)$ where h indicates the number of skeleton branches which, in turn, encompasses n grid points.

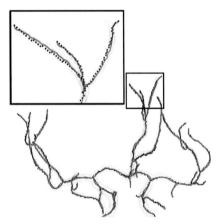

Figure 6.5: The figure shows two skeletons on different accuracy scales. Please notice that the dashed one is generated based on subvoxel precision. Closer inspecting these skeletal structures with respect to the vascular wall, it is obvious that the skeleton on voxel precision (solid line) suffers from inaccuracies caused by discretisation artefacts (cf. [VUB07]).

The overall idea behind this method is to detect a global point $\mathbf{p}^\star$ with the highest distance to the boundary. Afterwards, another interior point $\mathbf{p} \neq \mathbf{p}^\star$ is selected having the largest distance to $\mathbf{p}^\star$ which is then used as starting point of a skeleton branch. The branch by itself is established by performing a "gradient descent, back-tracking procedure on the fast marching time-crossing map" [VUB07]. This process is repeated until the distance between both points drops below a user-defined threshold. In the following, this procedure is going to be discussed in more detail:

Step 1: The segmented object and its corresponding sub-voxel precise distance field are passed as input to the algorithm. Although the authors propose a two step mechanism in [VUB07] to calculate this distance map, it does not matter which technique is finally employed as long as it fulfils the demands of their method. Thus, the suggested approach is replaced by the technique presented in [HF07a] which provides a higher accuracy of the resulting time-crossing map (cf. Section 2.2.6).

For the purpose of segmentation, the original work has employed an active contours method introduced by Caselles et al. [CKS97]. Even though the technique is finally substituted by the approach discussed in Section 6.2.1, it is worth knowing that this surrogate constitutes a derivative of [CKS97] by itself. Once the vascular structure is segmented, T can easily be generated by applying the FMM. Please notice that the remaining content uses the terms *distance* and *time crossing map* synonymously. Both concepts only differ in the way how they are interpreting the values of their outputs. Basically, they are referring to the same task that quantifies the relation between the grid and the seed points. In this work, the set of seed points is initially populated by the object's boundary. An exemplary result is illustrated in Figure 6.6.

Step 2: By taking the time-crossing map, two further quantities can be obtained, the *speed image* and the point inside the object with the largest distance to the boundary (the *global maximum*). Both magnitudes are essential for all subsequent steps. The speed image (or velocity, $v_\mathbf{x}$) was already introduced during the discussion of the level set method in Section 6.1.1 and embodies a popular instrument to control the evolution process of an interface. In this context, the speed image v is derived from T using the following equation:

$$v_\mathbf{x} = \begin{cases} \left(\frac{a_\mathbf{x}}{b}\right)^2 & \text{if } a_\mathbf{x} \in \Omega^+ \\ 0 & \text{otherwise} \end{cases}, \tag{6.36}$$

with $a_\mathbf{x} \in \Omega^+ \subseteq \mathbb{R}^n$ being the distance values covered by the object and $b = \max(T)$. The global maximum point is then detected by considering all peaks inside the distance map.

 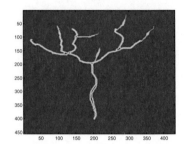

Figure 6.6: The figure illustrates an exemplary result of the FMM initialised by the object's contour. **(Left:)** The input image (taken from [Sps]). **(Right:)** The time-crossing map generated based on the FMM presented in [HF07a]. Please notice the outer part of the object has been set to zero in order to emphasise the interior of the object.

Therefore, T is sampled in scanline order with the aim to determine the first global maximum denoted by $\mathbf{p}^\star$. This point is shown in Figure 6.7 as a yellow dot depicting the location of $\mathbf{p}^\star$ inside the vascular structure. Although the implementation behind this procedure is straightforward, it guarantees that the global maximum point is going to be part of the final skeletal structure (cf. [VUB07]).

   Step 3:   With access to $\mathbf{p}^\star$, the authors are able to introduce the so-called *Augmented Fast Marching Method*. Together with the speed image v, $\mathbf{p}^\star$ is used to initialise the level set-driven evolution process (based on the fast marching approach). The augmentation of the FMM then calculates (in parallel) the piecewise linear geodesic distance from $\mathbf{p}^\star$ to all remaining grid voxels inside the object. Therefore, the smallest geodesic distance residing in one of the 26 adjacent neighbours is selected and subsequently increased by the distance to the currently selected grid location. Thus, a second distance map is generated carrying all geodesic distances emanating from $\mathbf{p}^\star$: $T^\diamond$. Once this evolution process is terminated, a point is selected and further assumed to be the end point of a skeleton branch. This is realised by searching that point having the highest geodesic distance in $T^\diamond$. If there is more than one point which share the same maximum distance, the location with the largest value in T is taken:

$$\mathbf{p}^\diamond = \operatorname*{argmax}_{\mathbf{p}_x} \; T(\mathbf{p}_x) \quad . \tag{6.37}$$

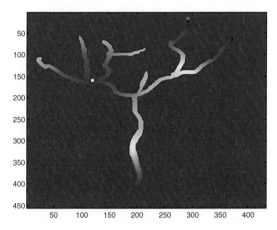

Figure 6.7: The figure illustrates the working principle of the augmented fast marching method. While the global maximum point is drawn in yellow inside the vascular structure, its furthest geodesic partner is depicted in pink.

Step 4: Given $\mathbf{p}^\star$ and $\mathbf{p}^\circ$, the task is now to find the centre line that represents the branch between $\mathbf{p}^\star$ and $\mathbf{p}^\circ$. According to [VUB07], this branch is found by solving the minimum-cost path problem:

$$\operatorname*{argmin}_{C^{\mathbf{p}^\star \to \mathbf{p}^\circ}} \int_0^N c(C(u))\,du \quad , \tag{6.38}$$

where $c(\cdot)$ indicates a cost function and $C(u) : [0, \infty) \to \mathbb{R}^n$ a parametrised path of length N inside the object. The goal is now to find that curve between $\mathbf{p}^\star$ and $\mathbf{p}^\circ$ among all other paths $C^{\mathbf{p}^\star \to \mathbf{p}^\circ}$ that minimises the cumulative costs as defined in (6.38). The authors Uitert and Bitter emphasise the use of the speed image as cost function ($c(\cdot) \widehat{=} v(\cdot)$) since it guarantees that no skeleton corners are hugged during the gradient descend back-tracking procedure.

> "Thus, the minimum path in the time-crossing map found during the back-tracking method will result in a branch that is in the central region of the object". [VUB07]

Step 5: The actual skeleton branch C is finally determined based on a gradient descent back-tracking approach starting at $\mathbf{p}^\circ = C(0)$. By following ∇T the goal is to reach the global maximum point $\mathbf{p}^\star$:

$$\frac{dC}{du} = -\frac{\nabla T}{\|\nabla T\|} \quad . \tag{6.39}$$

In more detail, the gradient descend procedure is operating on the local gradients which are generated based on the values in T. This can be realised, e.g. based on a linear interpolation scheme exploiting the eight gradients located at the corners of that voxel which currently accommodates the sample location (on sub-voxel precision): $\mathbf{p}^t$. Having $\mathbf{p}^t$, the next sample point $\mathbf{p}^{t+1}$ is calculated as follows:

$$\mathbf{p}^{t+1} = \mathbf{p}^t - \dot{\tau}\,\frac{(\hat{\nabla} T)}{\|\hat{\nabla} T\|} \quad , \tag{6.40}$$

where $\hat{\nabla} T$ depicts the local gradient. Moreover, the authors propose $\dot{\tau} = 0.01$ as suitable step size for determining the next sample location. Other sources[10] increase this value to 1.0 by coupling it with the *Runge-Kutta* method[11]. Unfortunately, both approaches have not been applicable in this project due to the problem illustrated in Figure 6.8 (also valid for 3D). Strictly speaking, the figure shows the back-tracking while it is lost after some iterations. It is supposed that numerical inaccuracies coupled with the huge step size of 1.0 are responsible for the identification of a wrong sample location sabotaging all subsequent iterations. In contrast to this, a step size of 0.01 requires an unacceptable amount of time for generating the skeleton.

Even if this problem does only occur rarely, its impact is enormous since any gap inside the skeletal structure prohibits the feature generation. Experiments showed that setting $\dot{\tau}$ to 0.5 in combination with the *Runge-Kutta* method is solving this issue. Nevertheless, the procedure reacts highly sensitive to changes of this parameter. Please recognise that the condition $\|\hat{\nabla} T\| \neq 0$ holds due to the fact that T is strictly monotonically increasing.

In addition to this, the FMM shows a singularity at $\mathbf{p}^\star$ inside the time-crossing map. This singularity is a by-product of the initialisation phase where $\mathbf{p}^\star$ has been exploited as seed for the evolution process. Thus the back-tracking procedure is not able to reach this destination natively. As a proper solution, the authors propose "to track the tracking" until it reaches $\mathbf{p}^\star$ in a distance less than one voxel unit. Afterwards, with the aim of bridging this gap, all pending sample locations (skeleton points) are obtained by a simple interpolation:

$$\mathbf{p}^{t+1} = \mathbf{p}^t - \dot{\tau}\left(\dot{\eta}\,\frac{(\hat{\nabla} T)}{\|\hat{\nabla} T\|} + (1 - \dot{\eta})\frac{\overline{\mathbf{p}}}{\|\overline{\mathbf{p}}\|}\right) \quad , \tag{6.41}$$

---

[10]http://www.mathworks.com/matlabcentral/fileexchange/24531-accurate-fast-marching, [online: 19th August 2015]

[11]http://mathworld.wolfram.com/Runge-KuttaMethod.html, [online: 19th August 2015]

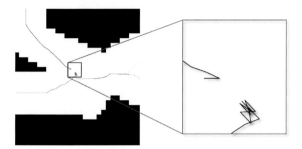

Figure 6.8: The figure illustrates the problem that appears during the back-tracking procedure when an inappropriate step size has been selected. (**Left:**) Zoomed junction segment of the vascular structure in 2D, where three skeleton branches do not join as expected. (**Right:**) Taking a closer look at this section, it is obvious that the back-tracking procedure took the wrong direction with the consequence that the algorithm is lost until a timeout signal terminates its execution.

where $\bar{\mathbf{p}}$ is the direction vector pointing from $\mathbf{p}^t$ to $\mathbf{p}^\star$. The interpolation weight $\hat{\eta}$ is defined based on the remaining distance to the global maximum point which is $\hat{\eta} = \|\bar{\mathbf{p}}\| / (\text{voxel size})$ (cf. [VUB07]). The branch tracking process is terminated if $|\mathbf{p}^t - \mathbf{p}^\star| \leq \hat{\tau}$.

Step 6:   After introducing the fundamental concept of extracting *sub-voxel precise skeletons*, the last paragraph is devoted to the final completion stage responsible for merging all branches to one single structure. Therefore, all possible centre lines have to be determined by repeating the previous steps. Once a new branch is determined, its sample locations are merged with the already existing ones: $\mathcal{K}^0 = \{\mathbf{p}^\star\}$, $\mathcal{K}^{t'+1} = \mathcal{K}^{t'} \cap C^{t'+1}$ with $C^{t'+1} = \{\mathbf{p}^0, \mathbf{p}^1, \dots \mathbf{p}^N\}$ and t' indicating the current iteration. Having $\mathcal{K}^{t'+1}$, a further branch is calculated by passing this set to the augmented FMM and a new $\mathbf{p}^{\diamond, t'}$ is determined representing the end point of the next skeleton branch. Restarting the gradient-descend back-tracking procedure at $\mathbf{p}^{\diamond, t'}$, the corresponding centre line can be extracted. Being aware of the singularities close to the seed points, the authors propose the same approach as introduced above. Each time the distance of the current sample location is less than one voxel unit, the pending points are approximated by Equation (6.41). According to the property of being strictly monotonic, the minima inside the time-crossing map are localised in the surrounding area of the previously identified skeleton points and thus, new branches are forced to intersect with this already existing structure. This process ends if the length of a new branch does not exceed a certain threshold μ. Therefore, the authors suggest to take the *maximum diameter* of the object which is approximated by taking two times the global maximum value (distance to the boundary). Hence, the algorithm stops its execution if the length of the new centre line drops below μ.

Please keep in mind that one of the major strengths of this method is its capability of determining the number of branches autonomously. It does not require any user input while the number of parameters to be set is negligible.

6.4 Abdominal Aorta Registration Using Two Opposing Matching Algorithms

Having in mind that the aorta structure has been segmented and shrunk to its centre line, the following content is dedicated to the actual matching process. This includes both the feature extraction and the detection of correspondences leading to minimum assignment costs. The latter is evaluated based on two different approaches which have been already introduced in the previous chapters. While the first is known from Chapter 3, the second one was deeply discussed in Chapter 5. Since the implementation details are already given in Section 2.4.2, this content part does only cover the parametrisation and the adaptations which have been applied in order to satisfy the demands of this project.

6.4.1   Intrinsic and Extrinsic Feature Extraction

As explained at the beginning, this section employs two opposing matching algorithms which, in turn, exploit their own features. The first set is calculated without taking into account external information and thus, it is referred to as *intrinsic features*. The second description type is determined with respect to the kidneys and the L4 segment of the lumbar spine, the *extrinsic features*.

Intrinsic Features   These features are extracted by exclusively considering the aorta skeleton and its geometry. Strictly speaking, the intrinsic description is implemented as a set of sequences encompassing the radii of maximum balls used as representation for the single skeleton branches. These radii are obtained by exploiting the sampling scheme introduced in Chapter 3, where the authors have scanned each skeleton path with a fixed number of equidistantly distributed points. However, instead of using maximum disks, maximum balls are fitted into the vascular body at each sample location. The use of 3D maximum balls is a popular instrument in context of extracting 3D skeletal structures for the task of localising skeleton points (cf. Chapter 2).

The actual calculation is straightforward and realised by utilising the time-crossing map T a further time in this project. By initialising the Fast Marching Method with the segmented vascular aorta wall, the radius of a certain maximum ball can be approximated just by taking the T-value at the position of its sample location. Please notice that this estimation is easier to implement than the actual skeleton extraction process. Instead of determining the best fitting ball with respect to the boundary of the object, the skeletal structure is already known and only the minimum distance to the boundary has to be recovered during the feature extraction process.

Extrinsic Features   The extrinsic descriptors incorporate external information for the purpose of representing the aorta centre line. Strictly speaking, the kidneys and the L4 segment of the lumbar spine are utilised to enrich the actual feature generation process. Please consult Section 6.2.2 to obtain further details about the segmentation of these organs.

In comparison both approaches are almost similar to each other and are sharing multiple properties. Basically, they are operating on the same sampling procedure that scans the shortest path $\rho(v, u)$ along the skeletal structure. According to Section 2.2.5, the shortest path is defined by two end points (or nodes) v and u having the shortest distance through the graph G. This path is sampled by a set of K equally distributed points. Subsequently, each sample point $\mathbf{p}_{i=1,\ldots,K}^{(v,u)}$ is expressed by a rotation and scale invariant identifier consisting of four angles in total which are calculated with regard to the kidneys ($\mathbf{e}_1, \mathbf{e}_2$) and the spine ($\hat{\mathbf{e}}$). Finally, each sample point is respectively connected by a vector with the L4 segment ($\mathbf{q}_{j=1,\ldots,K} = \mathbf{p}_{i=1,\ldots,K}^{(v,u)} - \hat{\mathbf{e}}$), in order to generate the actual descriptor: $\mathbf{x}_{j=1,\ldots,N} = [(\alpha^{(v,u)}, \beta^{(v,u)}, \gamma^{(v,u)}, \delta^{(v,u)})_{i=1,\ldots,K}]$ (with N being the number of shortest paths):

$$
\begin{aligned}
\alpha_i &= \arccos(\langle \hat{\mathbf{v}}_1, \hat{\mathbf{q}}_i \rangle) \\
\beta_i &= \arccos(\langle \hat{\mathbf{v}}_2, \hat{\mathbf{q}}_i \rangle) \\
\gamma_i &= \arccos(\langle \hat{\mathbf{n}}, \hat{\mathbf{q}}_i \rangle) \\
\delta_i &= g(\langle \hat{\mathbf{v}}_1, \hat{\mathbf{q}}_i \rangle, \langle \hat{\mathbf{v}}_2, \hat{\mathbf{q}}_i \rangle)
\end{aligned}
\quad , \tag{6.42}
$$

with $\hat{\mathbf{q}}_i$ being $\mathbf{q}_i$ on unit length and the unit vectors $\hat{\mathbf{v}}_1, \hat{\mathbf{v}}_2$ and $\hat{\mathbf{n}}$ correlating to:

$$
\mathbf{v}_1 = \mathbf{e}_1 - \hat{\mathbf{e}}, \qquad \mathbf{v}_2 = \mathbf{e}_2 - \hat{\mathbf{e}}, \qquad \mathbf{n} = \mathbf{v}_1 \times \mathbf{v}_2 \quad . \tag{6.43}
$$

Moreover, the function $g(\cdot, \cdot)$ in (6.42) has the following definition:

$$g(\mathbf{v},\mathbf{v}') = \begin{cases} \arctan(\mathbf{v}/\mathbf{v}') & \mathbf{v}' > 0 \\ \arctan(\mathbf{v}/\mathbf{v}') + \pi & \mathbf{v} \geq 0, \mathbf{v}' < 0 \\ \arctan(\mathbf{v}/\mathbf{v}') - \pi & \mathbf{v} < 0, \mathbf{v}' < 0 \\ +\pi/2 & \mathbf{v} > 0, \mathbf{v}' = 0 \\ -\pi/2 & \mathbf{v} < 0, \mathbf{v}' = 0 \end{cases} , \qquad (6.44)$$

Please notice that $\mathbf{v}_1, \mathbf{v}_2$ and $\mathbf{n}$ are linear independent by definition. Hence, they are forming a plane $\mathcal{A}$ as illustrated in Figure 6.9. This fact is also exploited during the matching process where the plane is recruited to establish the mutual exclusion (mutex) constraints.

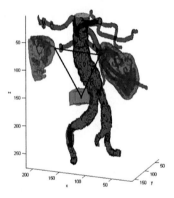

Figure 6.9: The figure indicates the plane $\mathcal{A}$ (depicted with black lines) which is spanned between the kidneys and the L4 segment of the lumbar spine (these organs are drawn in green). Moreover, the figure shows the segmented vascular structure (red) of the aorta.

6.4.2 Skeleton Matching for the Purpose of Registration

The finial registration performance strongly depends on the quality of the correspondences determined during the matching process. This result, in turn, relies on the fitness of the features and equally on the matching algorithm that is utilised to optimise the costs for aligning them. Strictly speaking, the features and the matching technique have to be chosen carefully in order to achieve accurate and robust results. Being aware of this correlation, two matching methods are involved for the task of evaluation (cf. Section 8.4), namely the Hungarian method and the Maximum Weight Cliques (MWC) approach. Both techniques are already known from the previous chapters and are explicitly described in Chapter 2.

Thus, the reader can refer to these content parts to obtain profound insights. This section is dedicated to the configuration details which are subject of the discussion below. In particular, the parametrisation of the MWC technique is comprehensively covered.

However, before the focus switches to the detection of MWC, the following lines are devoted to the overlapping part of both approaches, namely the input data in form of shortest path descriptors. In order to enable the matching algorithms to process this data, their path distances have to be calculated, respectively. Let $\rho(v,u)$ be a path in G between the nodes v and u and $\rho(v',u')$ a second one in G'. The actual path distance is obtained by:

$$d(\rho(v,u),\rho(v',u')) = \psi(\mathbf{x}^{(v,u)},\mathbf{x}^{(v',u')}) \quad , \tag{6.45}$$

where the $\mathbf{x}$ symbols are depicting the path signatures and $\psi(\cdot,\cdot)$ a time series matching function, e.g. the Dynamic Time Warping (DTW) (cf. Section 2.4.3.1). The DTW is an appropriate instrument for this task due to its capability of performing an elastic alignment between these two sequences. Taking this measure, all path signatures are transformed to path distance values with the intent to determine the matching costs between the nodes $v \in G$ and $v' \in G'$. This calculation is undertaken by the Optimal Subsequence Bijection (OSB) method (cf. Section 2.4.3.2). As a short reminder, the costs $c(v,v')$ are estimated by taking into account all paths emanating from v and v'. Therefore, the end points of both skeletons are traversed in the same order resulting in two series of path signatures which are subsequently processed by the OSB. Please bear in mind that the OSB is able to find a sub-sequence a' in a that best matches b' in b by skipping possible outliers (cf. [Lat+07b]).

Moreover, it is worth mentioning that the *extrinsic features* require a slightly more specific treatment with respect to their path signatures $\mathbf{x}^{(v,u)}$ and $\mathbf{x}^{(v',u')}$. Like in Chapter 5, the sequences are first separated into disjoint subsets before they are processed as follows:

$$d(\rho(v,u),\rho(v',u')) = \begin{aligned} &\psi(\mathbf{x}^{(v,u)}_{\alpha},\mathbf{x}^{(v',u')}_{\alpha}) + \psi(\mathbf{x}^{(v,u)}_{\beta},\mathbf{x}^{(v',u')}_{\beta}) + \\ &\psi(\mathbf{x}^{(v,u)}_{\gamma},\mathbf{x}^{(v',u')}_{\gamma}) + \psi(\mathbf{x}^{(v,u)}_{\delta},\mathbf{x}^{(v',u')}_{\delta}) \end{aligned} \quad . \tag{6.46}$$

Once all information is generated, they are passed to the Hungarian method, where the matching is performed based on a bipartite graph (cf. Section 2.4.2.1). The execution of the MWC approach requires a higher attention and is going to be discussed next.

6.4.2.1 Maximum Weight Cliques as Optimal Matching Configuration

The problem of determining MWC is already known from Section 2.4.2.2 and Chapter 5. Recalling that the matching problem is formulated as an integer quadratic program aiming at the identification of maximum weight cliques in an undirected affinity graph $G^{\star}$, it is obvious that further processing steps are required in addition to the previous ones. Moreover, it has to be considered that the final alignment has to satisfy a certain amount of mutex constraints.

In accordance to Chapter 5, the corresponding affinity matrix $\mathbf{A}$ is populated with *unary potentials* (matrix entries on the diagonal indicating the vertex weight) and *binary potentials* (off-diagonal entries representing the weights on the edges).

Unary Potentials   These quantities represent the weights of the vertices inside the affinity graph. With other words, they are indicating the similarity of the single assignments (v_i, v'_j) between two nodes $v_i \in G$ and $v'_j \in G'$, respectively. Moreover, these values are obtained in analogy to the PSSGM approach by utilising the OSB. It is worth knowing that the new matching method expects similarity values instead of dissimilarities in order to identify maximum weight cliques. This conversion is realised by passing the OSB outcome to a *Gaussian* distribution. The smaller the OSB distance, the higher the similarity. Furthermore, the *Gaussian* function can easily be parametrised for adapting its shape to the demands of the employed features. Please observe the parallels to Chapter 5 and to the original PSSGM.

Binary Potentials   According to the findings gathered in Chapter 5, the pairwise distance consistency is reused to generate the weights on the edges between two assignments: $r^\star = (v, v')$ and $r^\circ = (u, u')$ (with $v, u \in G$ and $v', u' \in G'$):

$$\mathbf{A}^{(r^\star, r^\circ)} = \exp\left(\frac{(f(v,u) - f(v',u'))^2}{2\theta^2}\right) \quad , \tag{6.47}$$

with $f(\cdot, \cdot)$ being a function that returns the Euclidean distance and θ, a parameter to adjust the influence of geometrical deformations. Enforced by the varying resolutions of the CTA series, all values are normalised in a range of $[0,1]$. Otherwise the approach would react sensitive to these different scales by falsely excluding correct assignments from the final configuration. This normalisation step is based on the height of the vascular structure.

Mutual Exclusion Constraints   The mutex constraints are useful in the light of reducing the search space while emphasising the binary potentials. Carefully chosen, they are able to drastically enhance the matching result in terms of accuracy, robustness and speed.

With the intention of registering two abdominal aorta structures, simple geometrical relations are exploited to fill the mutex matrix $\mathbf{M}$. Therefore, the plane $\mathcal{A}$ is recruited which has been introduced as a by-product during the feature generation process (cf. Section 6.4.1). Taking the vectors $\mathbf{v}_1, \mathbf{v}_2$ and $\mathbf{n}$, a local coordinate system is established capable of managing relations like *up/down* or *front/back*. Like Chapter 5, the actual implementation performs a coordinate-driven projection of each alignment pair (e.g. $r^\star = (v, v')$ and $r^\circ = (u, u')$). Subsequently, the differences between these projections are generated for each axis with the aim to express their relative positions to each other. In the following, the explanations are restricted to one dimension while the others are processed in the same way:

$$m_x = v_x - u_x$$
$$m'_x = v'_x - u'_x$$
$$\tag{6.48}$$

Only by monitoring the sign of both differences (m_x and m'_x), the algorithm includes ($\text{sign}(m_x) = \text{sign}(m'_x)$) or excludes ($\text{sign}(m_x) \neq \text{sign}(m'_x)$) an alignment from the final result. However, if the points of (v, u) or (v', u') are close to each other, the affected dimension is not considered due to the risk of measurement errors and inaccuracies.

As proven in Section 8.4, this kind of description performs robustly and achieves excellent results. An exemplary instance of two successfully registered aorta skeletons (based on the Maximum Weight Cliques approach) is illustrated in Figure 6.10.

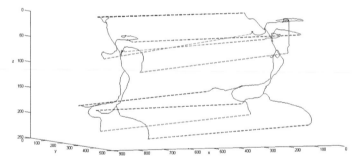

Figure 6.10: Exemplary matching result of two CTA series or rather their aorta skeletons which has been obtained by determining maximum weight cliques.

6.5 Summary

The content of this project introduced an appropriate approach to handle the assignment problem based on the task of aligning two abdominal aortas in analogy to the concept of the PSSGM. With the intention of registering pre- with post-operative data, a challenging but equally important task in context of professional medical diagnosis, treatment and after-care has been identified. As already mentioned at the beginning, aortic aneurysms (AAs) are a serious health problem even in developed countries and thus, they are worth of investigating in terms of new strategies to support today's physicians. The actual method has been evaluated on the basis of real CTA series affected by these AAs. Moreover, all evaluation results have been verified by a human expert qualified to rate the performance of the system towards the expectations of physicians on this pre- and post-operative data.

Given two CTA scans displaying the abdominal aorta of the same patient, one image contains the aneurysm (acquired beforehand) and the other one shows its absence after the surgery. The objective is to perform a registration of both vascular structures with the intent to allow a better assessment concerning the success of the medical intervention. Therefore, a bunch of complex methods are employed capable of extracting all information necessary to realise a robust and accurate registration system. Starting with two different segmentation approaches, the abdominal aorta is obtained by employing the technique presented in [Zha+08], whereas a fuzzy c-means clustering [KP07] is taken to extract the kidneys and the L4 segment of the lumbar spine. Once the segmentation is finalised, the vascular structure is passed to the next processing stage, namely the skeletonisation [VUB07]. Having the aorta's centre line, the feature generation process can start to generate two different features types: the intrinsic and extrinsic ones. They are obtained by sampling the shortest paths between the end points of the skeleton. While the intrinsic features are exclusively devoted to the aorta structure, the extrinsic ones involve the kidneys and the spine in order to derive four angles at each sample location. Finally, two graph-based matching algorithms are exploited to evaluate opposing registration configurations: The Hungarian method and the MWC approach [ML12] which are already known from the previous chapters.

Closely inspecting this processing line, several noteworthy points can be figured out. The sophisticated character of this method mainly arose by the simple composition of the aorta's overall tubular geometry. The less discriminative vessel structure together with a varying and rather modest resolution of the CTA scans, drastically increased the challenge of finding a proper way to distinguish different parts of the aorta. A fact that strongly conflicts with the goal to implement a robustly behaving matching method. Additionally, the task was exacerbated by the appearance or disappearance of vessel branches caused by a changing perspective on the abdominal aorta during its acquisition. Being aware of this, the use of topological features was not further investigated. Thus, it is even more impressive that the intrinsic features led to an overall acceptable registration result. This performance could be further enhanced by employing the extrinsic descriptors coupled with the MWC approach. However, please bear in mind that the processing of these stages requires a significant amount of computation time caused by the evolution-driven working principle of the segmentation and the skeletonisation.

In summary, the proposed technique has been able to establish almost all correspondence correctly. Moreover, the promising results encourage a further development of this technique. This includes the extension of the database as well as the improvement of the segmentation and the matching. For this purpose, it is planned to incorporate the DICOM positioning information in order to perform, e.g. a pre-alignment. Another idea is to use a mechanism for automatic medical image understanding as presented in [Tad07].

Chapter 7

3D Object Retrieval Based on Topological Features of 3D Curve Skeletons

As in the previous chapters of this thesis, the underlying sub-project is devoted to the task of object retrieval. In other words, it is dedicated to a deeper investigation of 3D curve skeletons in terms of generality and availability (cf. [Fei+13]). Therefore, a skeletonisation method is considered which has originally been proposed for the aim of object segmentation. Moreover, a set of objects is taken into account whose geometry is more related to surface skeletons. The outcome of this skeletonisation approach is then exploited in context of a similarity-driven measurement technique intended to be robust and accurate during the object retrieval process. Starting with the extraction of 3D curve skeletons (cf. Section 2.4.1), their discrimination power is then evaluated based on a set of topological oriented features.

As already mentioned, the actual skeleton extraction process has originally been proposed for the aim of object segmentation and is introduced by Reniers [Ren09]. Please keep in mind that the segmentation task does not require these strict limitations towards the curve skeleton as it is typically done by the object recognition task, e.g. a structure width of only one unit. Otherwise it would additionally exacerbate the goal of designing a robustly behaving feature set. To overcome these issues, the employed skeletonisation algorithm as well as the Dijkstra method are adapted to the given demands.

Although many research activities have risen over the last recent decades concerning this topic, the major amount was focused on the pure 3D skeleton extraction and only a little on the actual object recognition problem. Thus, this work shall further contribute to closing this gap by pushing the extraction closer to the recognition task which utilises the skeletons as basis for the actual similarity estimation between two 3D objects.

7.1   Fundamental Concepts

7.1.1   Feature Transform

The notion behind the feature transform is strongly connected to that of the distance one. While the latter determines the minimum distance between an arbitrary point $\mathbf{p}$ inside an object Ω to its closest boundary ($\partial\Omega$) location $\mathbf{x}$ (mathematically expressed as: $D(\mathbf{p}) = \text{argmin}_{\mathbf{p}} \|\mathbf{x} - \mathbf{p}\|$), the feature transform returns the actual point $\mathbf{x}$ instead of only its distance. Thus, the resulting feature set $\mathcal{F}(\mathbf{p})$ is defined as follows:

$$\mathcal{F}(\mathbf{p}) = \{\mathbf{x} \,|\, \mathbf{x} \in \partial\Omega : \|\mathbf{x} - \mathbf{p}\| = D(\mathbf{p})\} \quad , \tag{7.1}$$

where $\|\cdot\|$ depicts the ℓ^2-norm between two vectors (cf. [Ren09]).

Feature Points The term *feature point* might be rather misleading since other authors, e.g. in [KLT05; TVD08; GK09], are using this phrase as an indicator for shape locations on the object's boundary representing the most significant structural parts of it. They are commonly located on shape extremities having a high impact on the visual perception of the object. Especially in context of this work, they are frequently used as seed points for the computation of geodesics or even for the entire skeleton as demonstrated in Chapter 3. However, the following content interprets the term as those boundary spots with minimum distance to a certain point on the skeletal structure. Although such a relation is easily defined (cf. Equation (7.1)) for continuous shapes, their detection is far more challenging in discretised environments. Please recognise the correlation of this definition with the one which has been proposed by Blum 1967 concerning the skeletal structure [Blu67a].

7.1.2   The Cosine Similarity Measure

This section briefly introduces the measure selected for the purpose of estimating the similarity between two feature vectors based on the cosine of their opening angle. Thus, its name, *Cosine Similarity Measure*, is derived from its mathematical definition ($\psi_{\cos} : \mathbb{R}^n \times \mathbb{R}^n \to [-1, 1]$):

$$\psi_{\cos}(\mathbf{q}, \mathbf{q}') = \frac{\langle \mathbf{q}, \mathbf{q}' \rangle}{\|\mathbf{q}\| \cdot \|\mathbf{q}'\|} \quad , \tag{7.2}$$

where $\langle \cdot, \cdot \rangle$ depicts the dot product and $\| \cdot \|$ returns the length of a certain vector. In contrast to this formulation, its counterpart, the so-called *Cosine Distance*, is defined by a simple negation: $1 - \psi_{\cos}(\cdot, \cdot)$. Moreover, it is worth mentioning that this measure does not constitute a proper distance metric. The cosine similarity by itself belongs to the class of state-of-the-art techniques in that area. It is easy to implement, highly light-weighted and has an intuitive meaning. In addition to this, it performs well in many areas of computer science, e.g. in [Qia+04]. A further aspect supporting the use of this measure is its inherent normalisation scheme which natively produces an outcome in a range of $[-1, 1]$. Keep in mind that the comparison of objects based on their skeletal structures is realised by computing the cosine similarity between their corresponding feature vectors.

7.2 3D Object Skeletonisation Based on Jordan Curves

The skeletonisation approach proposed by Reniers has been designed in the light of 3D object segmentation [Ren09]. Although the method leads to promising results, the skeletal structure is suffering from being wider than one unit or rather voxel. While this structural anomaly does not affect the actual segmentation task, it provokes ambiguities during the feature generation process. Reniers justifies this degeneration of his skeletal structures by taking into account two different types of skeleton properties, namely the *intrinsic* and *extrinsic* ones. While the intrinsic attributes address the formal skeleton definition, the *extrinsic* properties are understood as application demands to process the given skeletal structures. Thus, it might happen that some of the intrinsic skeleton characteristics are neglected in favour of keeping or rather improving the *extrinsic* attributes.

In the scope of this project, the disregard of the *intrinsic* property which typically enforces the skeleton to have a width of one unit exacerbates the efficient computation of distinctive topological features. Strictly speaking, these imperfectly shaped structures are unattractive, e.g. for the detection of junction points. Knowing about the existence of other methods guaranteeing such a unit width, the stable topological outcome clearly endorses the employment of Reniers algorithm. Moreover, the proposed technique is capable of generating curve skeletons from a wide variety of arbitrary shaped 3D objects which usually tend to have a surface skeleton. However, in contrast to surface skeletons, the curve skeleton provides a convenient and higher intuitive access to its structure. Thus, it facilitates the extraction of topological information during the features generation process on the one hand. On the other hand it supports further progress that concerns the mapping of the PSSGM into other application domains. Hence, from a political point of view, this decision is reasonable since it encourages the internal objectives of the underlying thesis.

7.2.1   Tolerance-based Feature Transform

While Section 7.1.1 introduces the standard feature transform, the content below inducts the reader into the world of the so-called Tolerance-based Feature Transform (TFT). The TFT is an important instrument regarding the skeletonisation process which is going to be explained in the following. For the time being, just keep in mind that the method requires at least two feature points on the object's boundary ($\partial\Omega$) to be enabled to generate an appropriate curve skeleton. In contrast to this, the classical feature transform typically delivers only the closest boundary point for each $\mathbf{p}$ inside the object (Ω). To retrieve more than one boundary point having almost the same minimum distance to $\mathbf{p}$, the author suggests adding a user-adjustable tolerance parameter ϵ to Equation (7.1) with the intent to detect all points whose distances are close to the minimum one:

$$\mathcal{F}(\mathbf{p}) = \{\mathbf{x} \mid \mathbf{x} \in \partial\Omega : \|\mathbf{x} - \mathbf{p}\| \leq D(\mathbf{p}) + \epsilon\} \quad . \tag{7.3}$$

Moreover, the TFT alleviates the influence of discretisation artefacts as illustrated in Figure 7.1. As shown there, the TFT returns a better sampling result in the sense of approximating the boundary segment more accurately. Figure 7.1 demonstrates this by mapping a continuous curve into a discretised space. Subsequently, the continuously shaped boundary is processed exemplarily by employing the TFT with zero tolerance ($\epsilon = 0.0$) which behaves similar to the classical feature transform. In comparison to this, the TFT is also applied to its discretised version, but instead of using only one tolerance value, the transformation is respectively invoked by zero ($\epsilon = 0.0$) and one ($\epsilon = 1.0$). Even though the zero tolerance-driven approach performs successfully on the continuous shape, its discretisation requires a value of one ($\epsilon = 1.0$) in order to return all points on the curve and not only a subset of it ($\epsilon = 0.0$). Please consult [RT06] to retrieve deeper insights concerning the method's implementation.

7.2.2   Skeleton Extraction by Determining Jordan Curves

The actual skeletonisation method has been proposed for generating both multi-scale curve and surface skeletons. Its implementation utilises the Boundary-Distance Measure (BDM) and maps it into 3D space coupled with a simple thresholding strategy. However, due to the fact that this project exclusively focuses on the topological structure of an object and that the thesis by itself is dedicated to a deeper investigation of the PSSGM, the following discussion is limited to the topic of curve skeletons. Thus, the symbol $\mathcal{S}^{3,1}$ which typically indicates this skeleton type, can be reduced to $\mathcal{S}$ (cf. Section 2.4.1).

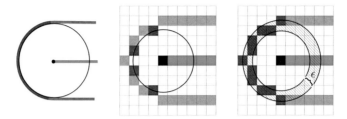

Figure 7.1: The figure illustrates the working principle of the TFT (cf. [Ren09]). It clearly shows the advantages of using an adjustable tolerance parameter (ϵ) in context of discretised structures. The TFT returns an optimal approximation of the half-circle. **(Left:)** Continuously shaped boundary part (grey line) and its sampling result without tolerance value (red) based on the outer skeleton point (black dot). **(Centre:)** Same result but in discretised space. **(Right:)** Optimal approximation using a tolerance value: $\epsilon = 1.0$.

Boundary-Distance Measure This measure was first introduced by Ogniewicz as *potential residual* [Ogn94]. It aims at the task of skeleton pruning with the concept of assigning a so-called importance value to each skeleton point. In order to realise this, two feature points (residing on the object's boundary) are selected with minimum distance to a certain skeleton point. Subsequently, at these locations the entire boundary is sliced into two sub-curves while the shorter one is returned as importance indicator. Please notice that this measure has a global character. In other words, all boundary points are considered at least once during the entire process. Moreover, the magnitudes of the values rise monotonically the more boundary points are encompassed by the shorter shape segment. In consequence of this, the measure has only one maximum peak among its set of all importance values.

Unfortunately, the BDM cannot natively be mapped into 3D. In order to tackle this issue and to process the 2D boundary of a 3D object in analogy to the working principle of the BDM, Reniers proposes splitting the boundary's surface into two 2D components for the generic case (and even for non-generic instances). Please consult [Ren09] in order to obtain a more detailed description. With the intent to introduce at least the core idea behind this methodology, a brief explanation is given for generic cases in the following. According to Reniers' thesis, let x_i and x_j be two feature points which are assigned to a certain skeleton point p with $i \neq j$. Having this information, the algorithm establishes a connection between x_i and x_j by estimating geodesics on the boundary's surface. In addition to this, further assume that Γ indicates the set of all geodesics available in the features space of an arbitrary point p inside the object:

$$\Gamma(\mathbf{p}) = \bigcup_{\mathbf{x}_i, \mathbf{x}_j \in \mathcal{F}(\mathbf{p})} \rho(\mathbf{x}_i, \mathbf{x}_j) \quad , \qquad (7.4)$$

with $\rho(\cdot, \cdot)$ being a function that takes two arguments for which all shortest path are returned connecting these two points. It is worth mentioning that $\rho(\cdot, \cdot)$ delivers multiple geodesics if the object has a symmetric geometry leading to more than one path sharing the same minimum length.

The actual classification of whether or not an object point belongs to the skeletal structure, can easily be performed by monitoring the size of Γ. Attributed to the symmetry property of a skeleton, its members embody at least two paths forming some kind of ring around the object known as *Jordan curve*. Hence, according to the formal definition, a point $\mathbf{p}$ belongs to the curve skeleton $\mathcal{S}$ if it establishes a Jordan curve:

$$\mathbf{p} \in \mathcal{S} \Leftrightarrow |\Gamma| \geq 2 \quad , \qquad (7.5)$$

Moreover, the concept of Jordan curves facilitates the introduction of a new importance measure. It straightly operates on the object's boundary. For every $\mathbf{p}_i \in \mathcal{S}$, the corresponding Jordan curve(s) is used to slice the 3D geometry into at least two components where the paths are indicating the corresponding cutting lines. For each component its surface size is computed and the smaller one is used to express the importance of a point inside the object. Mathematically speaking, a function $\psi(\cdot)$ exists which returns this magnitude for every object point by exploiting the smaller surface area:

$$\psi(\mathbf{p}) = \|\partial\Omega\| - \operatorname*{argmax}_{h \in \mathbb{D}} g(h) \quad , \qquad (7.6)$$

where $\|\partial\Omega\|$ indicates the size of the entire boundary. In contrast to this, the function $g(\cdot)$ does only return the specific area of an arbitrary surface part that has been collected in $\mathbb{D}$ during the slicing process. Please remember that the length of $\mathbb{D}$ derives from the number of elements in Γ. The inverse notation is necessary to react appropriately to non-generic cases characterised by $\|\mathbb{D}\| = \|\Gamma\| \geq 3$. The final pruning or rather the rejection of false positive points is then realised by selecting an appropriate threshold which implies the lower bound of importance for being a valid skeleton point.

7.2.3 Skeleton Extraction on Discretised 3D Objects

After discussing the extraction of curve skeletons in theory, this section closes the gap between the concept and its actual implementation. As mentioned in Section 7.2.1, the

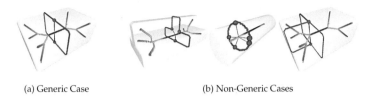

(a) Generic Case   (b) Non-Generic Cases

Figure 7.2: The figure illustrates some examples which represent either the generic case or the non-generic one (cf. [Ren09])

TFT constitutes an adaptation of the classical approach and provides a better sampling performance on discretised structures. However, the TFT by itself is not able to compensate all discretisation artefacts as they might occur in correlation with an even number of grid points between two boundary segments. Closely inspecting Figure 7.3a discovers that there is no *real* centre location that has the same minimum distance to both boundary fragments. In consequence of this, the method is disabled to determine the desired Jordan curve and thus, it loses its function of classifying object points as skeleton members. Figure 7.3a illustrates both the problem of finding the centre point and its related issue of determining (at least) two geodesics at minimum distance. Although the example is shown in 2D based on the features $\hat{x}$ and $\hat{x}'$ and a given object point $\mathbf{p}$, it stays valid in 3D space. Please observe that the 1D boundary illustrated in Figure 7.3a can also be considered as a single slice of an arbitrary 3D object.

A possible solution to this problem might be the introduction of a sub-unit resolution as proposed in Section 6.3. However, in context of [Ren09], another approach has finally been selected which utilises the concept of the TFT, namely the so-called Extended Feature Transform (EFT). The benefit of using this modified version is its insensitive behaviour towards the previously stated obstacles which are caused by discretisation. Moreover, the EFT is capable of reducing the amount of features points with the effect of an accelerated computation performance. The working principle of the EFT is straightforward and only incorporates adjacent grid points additionally to the feature detection process (cf. Figure 7.3b):

$$\mathcal{F}'(\mathbf{p}) = \bigcup_{x,y\in\{0,1\}\times\{0,1\}} \mathcal{F}((p_x + x, p_y + y)^\top) \quad . \tag{7.7}$$

By considering grid points in the surrounding area of $\mathbf{p}$, the problem of finding features on the object's boundary is solved. Connected to this, (at least) two geodesics can be localised for the purpose of classifying skeleton points. Figure 7.3b shows the shortest paths

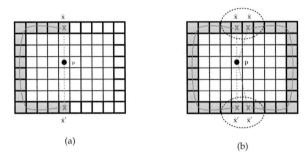

(a)

(b)

Figure 7.3: The figure illustrates the problem of an evenly sized grid structure accommodated between two boundary fragments. **(Left:)** No centre point can be detected with at least two feature points at minimum distance. In consequence of this, the estimation of two geodesics fails. In absence of theses shortest paths, no Jordan curve can be established and thus, the classification of skeleton points is disabled. **(Right:)** The solution to this issue realised by employing the Extended Feature Transform, a derivative of the TFT.

with identical lengths which could be established between $\{\hat{\mathbf{x}}, \hat{\mathbf{x}}'\}$ and $\{\breve{\mathbf{x}}, \breve{\mathbf{x}}'\}$, respectively. Thus, a Jordan curve is obtained that confirms the affiliation of $\mathbf{p}$ to the curve skeleton: $\mathbf{p} \in \mathcal{S}$. Moreover, it is worth mentioning that all geodesics are determined by employing the Dijkstra algorithm. However, the actual detection of Jordan curves is challenging since the curve might be a composition of multiple geodesics which occur if $\mathcal{F}'(\mathbf{p})$ consists of more than two features: $\|\mathcal{F}'(\mathbf{p})\| > 2$. Reniers tackles this issue by utilising an additional dilatation step with the aim to merge all sub-paths into one single construct. In other words, the morphological operation inflates the curves to a kind of *band*, the *Jordan band*. Finally, the borders of each Jordan band are traced in order to identify two or more connected border segments which endorse the existence of a Jordan curve [Ren09].

7.2.4   Quality Assessment of the Skeletal Structure

This section briefly discusses the observations collected during the implementation phase of Reniers' skeletonisation method. These findings are crucial with regard to further research activities since they indicate the prospects and limitations of the intended goal. Thus, being aware of them allows the definition of supplementary tasks which are necessary to reach the goal of recognising objects. The most important property that has been confirmed by the evaluation concerns the fact that the skeletons behave robustly in terms of their structures even under isometric transformations. This invariance had a general character for all objects which had been given to the system. Moreover, the resulting structures are thin, but do not fulfil the *intrinsic* width attribute of curve skeletons as defined in Section 2.4.1. Strictly speaking, it might occur that the skeleton branches are wider than one grid unit. These structural anomalies are the result of the EFT which incorporates additional object points in order to obtain the required boundary features. Although the problem seems to be trivial, it has a negative influence on the detection of the end and the junction points towards the skeleton since it leads to ambiguities of the skeletal structure and hence, their localisation becomes more complex. A further issue which is related to this arises during the computation of the distances between these points, e.g. end-to-junction or end-to-end point. In more detail, both point types have to be accurately detected to successfully suppress the emergence of false positives which could drastically affect the matching result.

A further obstacle of the skeletal structure concerns its connectivity since the skeleton branches are not fully connected to each other. This problem has to be ascribed to the importance measure running the risk of oscillating around the pre-defined threshold with the effect of generating unconnected skeleton points. However, these falsely classified skeleton parts do only exhibit low importance magnitudes, a fact that can be exploited for the application of multiple strategies to remove these unwanted curves, e.g. by discarding the entire branch or (in case of a higher importance) to close the gap with the intent to establish its connection.

These observations are already mentioned partially in [Ren09]. Nevertheless, they have to receive a higher attention in form of modifications which are going to be introduced next. These adaptations are dedicated to the enhancement of the robustness and the discrimination power of the feature set which additionally improves the overall recognition performance of the system. Please keep in mind that the term *feature* is used ambiguously in this project since it refers to the boundary points on the one hand and to the sequences of numbers which represent the characteristics of the underlying skeleton on the other.

7.3   Topological-Based Feature Set by Curve Skeletons

The skeletonisation process generates topologically robust curve skeletons, whereas the objective of recognition is facing the problem of inaccurately classified skeleton parts (cf. Section 7.2.4). In consequence of this, distances between these points cannot be calculated uniquely. Although a rough indication of their positions is given during the skeletonisation process, its precision does not fulfil the requirement of the feature generation process. However, if the junction and the end points are determined appropriately, the actual feature extraction process can be invoked. The resulting descriptors are designed to be capable of encoding the topological structure of the skeleton. This special view on the skeleton allows the algorithm to encode global information about the object which typically behaves insensitively to noise. Even though the topology remains mostly unaffected in presence of noise, it inherits the potential risk of being distorted during the skeleton pruning by falsely removed or kept branches. While geometrical-driven features are equally affected by, e.g missing skeleton branches, the impact of noise is higher on this data since deformations of the skeletal structure cannot be re-aligned by the skeleton pruning process.

Furthermore, the topology of a skeleton has further attractive properties regarding articulating objects since the arrangement of skeleton points stays almost the same in presence of such structural changes. Thus, a topological oriented feature set does still allow a proper adaptation of these deformations without any additional operations. Hence, the feature vector is primarily designed to cover the following aspects: First, only the topological data of the skeletal structure shall be incorporated without considering any geometrical information. Second, the feature set has to be invariant to rotation, scaling and translation (RST) in order to decrease the complexity of comparing two objects in 3D. Although topological data constitutes a powerful and highly flexible instrument in many situations, it is worth mentioning that the exclusive use of this data can lead to ambiguities like in the case of a horse and a dog or even a cat since all of them exhibit a highly similar alignment of skeleton branches. According to this, the project investigates the degree of discrimination that can be achieved by only considering the object's topology. The question itself is also highly interesting from a scientific point of view, especially in the light of articulating object parts. The actual feature space has five dimensions and is defined as follows:

$$q = (q_1, q_2, q_3, q_4, q_5)^\mathsf{T} \quad . \tag{7.8}$$

Feature 1   This quantity is determined by counting the number of skeleton end points $v_0^\star, v_1^\star, \ldots, v_K^\star$. It is depicted with q_1 and can be interpreted as an indicator for the shape complexity of an arbitrary 3D object.

Feature 2   The second descriptor (q_2) exploits the set of skeleton junction points $v_0^\circ, v_1^\circ, \ldots, v_N^\circ$. It can be considered as an extension of the complexity feature stored in q_1.

Feature 3   The third feature, denoted by q_3, determines the average number of outgoing paths emanating from a single junction point v_i°. It is modelled as the ratio of end to junction points accommodated by the skeleton. The higher the value, the more end points are connected to a single junction point on average:

$$q_3 = \frac{1}{q_2} \sum_{i=1}^{q_2} \sum_{j=1}^{q_1} f(\mathbf{v}_i^{\circ}, \mathbf{v}_j^{\star}) \quad , \tag{7.9}$$

with $f(\cdot,\cdot)$ being a binary function returning one if $\mathbf{v}_j^{\star}$ is connected to $\mathbf{v}_i^{\circ}$.

Feature 4 The magnitude of the fourth feature value indicates the average distance between all skeleton end points $(\mathbf{v}_i^{\star}, \mathbf{v}_j^{\star})$ with $j \neq k$. Therefore, the lengths of all shortest paths are taken into account by considering every combination of end points. Finally, this data is compressed to a single number that is stored in q_4:

$$q_4 = \frac{2 \cdot (q_1 - 1)}{q_1} \sum_{i=1}^{q_1} \sum_{j=i+1}^{q_1} l(\rho(\mathbf{v}_i^{\star}, \mathbf{v}_j^{\star})) \quad , \tag{7.10}$$

where $\rho(\cdot,\cdot)$ returns the shortest path between two end points, whereas $l(\cdot)$ calculates the length of this given path $(\rho(\cdot,\cdot))$.

Feature 5 The last feature covers the standard deviation of all shortest path lengths. It expresses the regularity of the different object parts related to each other. This information is carried by the last element (q_5) of the feature vector $\mathbf{q}$. Please notice that this deviation has been calculated in cooperation with the previously determined average of these path lengths (q_4):

$$q_5 = \frac{2 \cdot (q_1 - 1)}{q_1} \sum_{i=1}^{q_1} \sum_{j=i+1}^{q_1} (l(\rho(\mathbf{v}_i^{\star}, \mathbf{v}_j^{\star})) - q_4)^2 \quad . \tag{7.11}$$

7.4   Skeleton Adaptations for the Purpose of Feature Generation

After introducing a topological oriented feature set, this section approaches the topic of modifying the skeletal structure with the intent to retrieve the quantities as introduced in Section 7.3. These adaptations encompass the removal of outliers and the adherence of the *intrinsic* property enforcing a unit width of the structure as stated in Section 2.4.1. Hence, substantial variations between the skeleton and its general intrinsic properties shall be suppressed in order to meet the demands of the feature generation process.

In the following, a modified Dijkstra method is going to be proposed capable of removing outliers inside the set of skeleton points $p \in S$ without affecting the skeletal structure's consistency. In addition to this, it tackles the problem of partially occurring skeleton parts being wider than one grid unit. Particularly in view of the junction points, it might occur that they are not recognised by the native Dijkstra algorithm. Being aware of this issue, so-called skeleton nodal areas (SNAs) are incorporated with the aim to collect all $\mathbf{p} \in S$ which are representing the same junction point. Subsequently, this collection is considered as a single junction node inside the skeleton graph. Altogether, the proposed method consists of multiple stages which are going to be summarised below:

Step 1:   The skeleton structure is transformed into a connected, undirected skeleton graph which takes every skeleton point as a single node. Having a number of N skeleton points, the graph encompasses N skeleton nodes which are only connected if they are sharing at least one voxel edge. The weights of each link are then defined as the distance between the corresponding skeleton points, respectively. By re-arranging these weights to a matrix, a $N \times N$ symmetric *adjacency matrix* $\mathbf{A}$ is obtained.

Step 2:   The skeleton nodes are then checked in terms of being a member of a so-called skeleton nodal areas (SNAs). This is implemented by taking into account only those nodes ($\hat{e}_i$) which exhibit more than two adjacent neighbours. Finally, a $3 \times 3 \times 3$ cube is utilised as structure element gathering all $\hat{e}_i$ which are adjacent to each other forming a SNAs, respectively. The final collection of nodes is then labelled as a single junction node in that skeleton. This way, the method suppresses the skipping of junctions in the skeletal structure. In addition to this, the merging is required due to the risk of generating false positives among the set of valid junction points.

Step 3:   Once all SNAs have been determined, the procedure starts to compute the shortest paths between all pairs of end points $((\mathbf{v}_i^\star, \mathbf{v}_j^\star)$, with $i \neq j)$. This is realised by exploiting the modified Dijkstra algorithm which is applied to the previously generated skeleton graph. In this context, it is worth mentioning that each end point $\mathbf{v}_i^\star$ of the skeleton is assigned to its first SNA crossed by the corresponding path on its way to the destination node $\mathbf{v}_j^\star$ during this process.

Step 4:   Finally, the resulting Dijkstra paths are validated according to two exclusion criteria in order to remove misclassified end points, a topic which has not been considered yet. Strictly speaking, an end point is removed from the skeletal structure when at least one of the following constraints is fulfilled:

- The selected end point resides between two skeleton nodal areas.

- The end point is part of a cluster consisting of multiple end points which are connected to the same SNA. Assuming that $\mathbf{v}_i^\star$ and $\mathbf{v}_j^\star$ are two different end points ($i \neq j$) inside the same skeleton graph and that their shortest path $\rho(\mathbf{v}_i^\star, \mathbf{v}_j^\star)$ does not pass any SNA, only one of them is kept. Therefore, the point with the largest path length to the jointly used SNA is selected while all remaining cluster members are removed from the skeleton.

Figure 7.4 illustrates both cases. Please observe that the points $\mathbf{v}_2^\star$ and $\mathbf{v}_3^\star$ are removed from the skeleton since they are matching one of the previously stated exclusion criteria. Using this approach, outlier elements can be excluded elegantly from the resulting skeleton structure and do not distort the feature generation process.

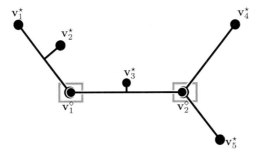

Figure 7.4: The figure illustrates the topology of an arbitrary skeleton consisting of five end points and two skeleton nodal areas. Moreover, the structure is suffering from two outliers, namely $\mathbf{v}_2^\star$ and $\mathbf{v}_3^\star$. Please notice that these falsely detected points are going to be removed during the execution of the proposed post-processing step since they are matching respectively one of the two exclusion criteria.

Step 5:   The last step is only mentioned for the sake of completeness and encompasses the computation of all feature values as they have been introduced in Section 7.3.

7.5   Summary

This project has been devoted to a deeper investigation of curve skeletons derived from 3D objects. The extraction of reliable curve skeletons is crucial in context of using the Path Similarity Skeleton Graph Matching as introduced in Chapter 3. Although the project is primarily dedicated to the overall objective of the present thesis, the content has additionally been instructed to acknowledge the topology of a skeleton. Moreover, possible synergies between both feature types, namely the topological and the geometrical ones, should be identified.

In contrast to Chapter 5, in which the depth data has been recorded from a fixed camera position, this project exploits the entire 3D object given as a voxelised input structure. With access to the complete boundary of an object, the possibility of extracting a skeletal shape is provided. While Chapter 6 also extracts curve skeletons form its input, the content above is aiming at different tasks: (i) The extraction of curve skeletons from objects which rather support the creation of surface skeletons and (ii) the exclusive use of skeleton descriptors. Strictly speaking, topological characteristics are employed to populate a five-dimensional feature vector. This approach is highly beneficial for this work since the other chapters have primarily been assigned to further aspects like boundary-based properties or externally defined feature generation processes. Moreover, the topology of a skeleton has a lower sensitivity to noise than the geometrical features which characterise the object's boundary.

For the purpose of generating the desired curve skeletons, a method is employed which has originally been proposed by Dennie Reniers in [Ren09]. The core idea behind this technique utilises the 2D Boundary-Distance Measure by mapping its concept into 3D space. Having this new measure, every 3D object point can be assigned to an importance value by taking the entire boundary of the object into account. In order to compute this importance indicator, at least two feature points have to be localised on the boundary with the intention of generating so-called Jordan curves. Using these curves as cutting lines, the object is sliced into different components to define the actual importance measure as the boundary size of the smallest partition. The final classification is then established by introducing a threshold which is applied to this importance data. Please notice that this approach has originally been developed with the aim to segment 3D objects instead of recognising them.

Being able to generate curve skeletons as described in [Ren09], their structural quality has been inspected in the light of object recognition or rather in context of generating a distinctive feature set. The results of this analysis discovered obstacles exacerbating the extraction of meaningful descriptors: First, the skeletal structures are allowed to be wider than one grid unit acting contrary to the detection of both the junction and the end points of the skeleton. Second, the threshold-guided skeleton pruning runs the risk of accidentally removing valid skeleton points with the result of disjoining randomly branches from the global structure. The deletion or annexation of these parts can be solved heuristically by exploiting a workaround that examines the curves in terms of length and importance. Contrary to this, the feature generation issue requires more attention to be tackled appropriately. Therefore, a modified Dijkstra algorithm has been proposed capable of eliminating end point outliers while emphasising the junctions by merging ambiguous skeleton regions to one group which is then considered as a single node.

Although the evaluation of the proposed technique has only achieved moderate retrieval results, the more important task, namely the extraction of curve skeletons, performed quite well. However, as already expected at the beginning of this project, the topological information is not sufficient to provide highly discriminative properties compared to other features, e.g. the geometrical ones. The explanation for this strongly relates to the level of detail that is transported by the boundary. Moreover, the method is suffering from conceptual problems: First, the database used for evaluation is highly unbalanced and did not always provide a second instance for each object class. Thus, the algorithm was not able to find a *real* counterpart for every query. Second, the heuristic approach responsible for the creation of the document result lists did not perform well in all cases. Indeed, the method delivered quite promising results related to the human cognition in terms of rating the similarity between 3D objects. Thus, the future work is going to investigate the following aspects: First, the method will be improved with regard to its database and its mechanism to determine a proper set of result documents regarding both the retrieval and the ground truth. Besides this, the approach of Dennie Reniers will be coupled with the idea of using a sub-unit resolution to find more appropriate centre locations inside an object. This shall suppress the emergence of outlier elements caused by the use of the Extended Feature Transform. Further improvements are concerning the topological oriented feature set and the application of surface skeletons. The merging of topological and geometrical feature is expected to increase the potential of better distinguishing different object parts.

Chapter 8

Experiments and Results

This part of the thesis collects the results which have been obtained during a thoroughly planned evaluation of each single project that has been introduced above. It further provides information about configuration values towards the parameters which have been applied in the following. In addition to this discussion, an assessment is respectively given that analyses the underlying recognition performance in terms of strengths and weaknesses.

8.1 Chapter 03: 2D Object Retrieval Based on 2D Skeletons

At first the recognition performance of the Path Similarity Skeleton Graph Matching (PSSGM) is going to be discussed in detail. The technique is introduced in Chapter 3 according to [BL08] and reused at several stages in this work, e.g. in Chapter 5 and Chapter 6. However, before these projects were initiated, the PSSGM had deeply been studied to retrieve a realistic assessment of its feasibility. Therefore, the method has been re-implemented in [Hed+13] with the intention of analysing it.

8.1.1 Evaluation Setup

The actual evaluation is performed on top of three databases, namely *Aslan and Tari* [AT05], *Kimia-99* [SKK01] and *Kimia-216* [SKK04]. All shape instances are reduced to their skeletal structures by the Discrete Curve Evolution (DCE) (cf. Section 3.1.1). Therefore, the open code[1] published by Xiang Bai is used which combines the DCE with a skeleton growing approach that is limited to a maximum number of 15 vertices. With access to resulting the skeleton branches, the *path distance* is calculated by scanning each shortest path based on a fixed number of 50 equally distributed sample points. The merging of the radii data with

---

[1]`https://sites.google.com/site/xiangbai`, [online: 19th August 2015]

the actual path lengths information is then performed by weighting the latter as proposed by the authors Bai and Latecki ($\eta = 40$). Starting with this configuration, the retrieval results are generated for all databases. In other words, each instance is selected once as query and subsequently applied to all remaining shapes. The outcome of this operation is finally arranged in descending order according to the similarity values between the query and the targets as illustrated in Table 8.1.

2	2	1	1	0	0	0

Table 8.1: The table illustrates two exemplary outputs of the retrieval system utilised to rate its recognition performance. By passing an arbitrary query (here a cat and an elephant), an object list is returned with those targets having the highest shape similarity to the query arranged in descending order (cf. [Hed+13]).

Finally, each column is consolidated by counting each true positive as one and each false positive as zero. If the column sum equals the number of objects in the database, all targets at this position have been determined correctly and belong to the category of the query. Hence, the matching result is perfect for this rank. The goal, of course, is to increase the amount of positive hits in each column close to this maximum.

8.1.2 Evaluation Results

The following content discusses the results which have been achieved by applying the re-implementation of the PSSGM to the shapes taken from *Aslan and Tari*, *Kimia-99* and *Kimia-216*. Therefore, the matching quality is analysed in terms of the similarity raked output list, the corresponding average precision and its precision/recall curve.

Aslan and Tari   Taking the configuration stated above, the method achieves an overall precision of 0.947 based on the *Aslan and Tari* data set. This magnitude relates to the total number of hits which have been monitored in the first three columns of the similarity ranking returned by the algorithm, namely [43 | 41 | 41]. Please notice that [44 | 44 | 44] denotes the optimum towards eleven object classes consisting of four instances, respectively. Although the method performs quite well on these shapes, it should be noted that the crocodile class was the most difficult one to be matched. Closely inspecting the mismatches displayed in Table 8.2, one observes that all crocodile instances are corrupted by false positives. In addition to this, it also reveals that a turtle has been considered as the most similar object to one of the crocodiles.

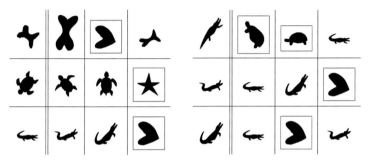

Table 8.2: The table provides an outline of the retrieval result with all mismatches that have occurred by applying the re-implemented PSSGM to the shapes of Aslan and Tari. Please observe that all crocodile instances are corrupted by outliers (cf. [Hed+13]).

A deeper analysis discovered that all crocodile shapes are suffering from *spurious skeleton branches*. Coupled with the *one-to-one matching* caused by the Hungarian method, wrong assignments have been established with the result of increasing matching costs. Consequently, the similarity between the two instances of the crocodile class dropped drastically. Moreover, it is interesting to see that these spurious branches especially emerged inside the crocodiles' tails having a strong articulation. This assumption is confirmed by manually pruning these violating structures with the effect of an enhanced recognition performance: [44 | 43 | 41] (equals an average precision of 0.97). This observation clearly shows that the method is extremely sensitive to spurious branches capable of affecting the overall shape similarity and thus, the overall retrieval performance.

Kimia-99   In addition to the Aslan and Tari shapes, the Kimia-99 database was chosen as a further evaluation set in order to retrieve a more meaningful analysis. The selected shapes encompass six object classes with ten instances for each category. A short outline of these objects is given in Table 8.3. This composition of object classes is highly heterogeneous and contains both rigid and articulating instances. Besides this, the shapes are suffering from different degrees of distortions. The recognition performance is measured in the same manner as before. Additionally, a precision/recall curve is provided in Figure 8.1 which

Table 8.3: The table displays some of the object categories taken from the Kimia-99 database for the purpose of evaluation. As illustrated, the subset encompasses both rigid and non-rigid objects. Moreover, some of the objects are heavily distorted in their shape.

precisely captures the evolution of the precision relative to the number of result documents.

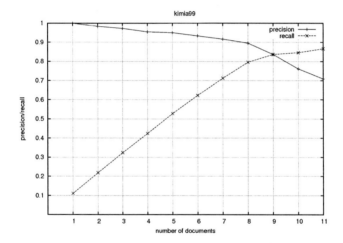

Figure 8.1: The figure shows the evolution of the precision and the recall according to an increasing number of target objects that have been considered at each evolution step. As input for the PSSGM a subset of 60 objects has been used taken from the kimia-99 database.

Roughly speaking, the method achieves an average precision of 0.837 which is derived from the similarity-driven outcome: [60 | 58 | 57 | 54 | 56 | 51 | 49 | 45 | 22]. The list nicely reflects the curve which is shown in Figure 8.1, where the precision directly starts to drop slowly with an increasing amount of documents. Please further recognise that the limited view on the first nine elements does not change the observation although the result list could be extended to eleven. However, even in presence of this challenging set of objects, the PSSGM is able to obtain an acceptable precision that is very promising regarding further research activities. A deeper analysis of all mismatches additionally confirmed that occlusions and heavily distorted object parts have been responsible for the generation of falsely established assignments decreasing the similarity between two objects. It is worth mentioning that spurious branches constituted the second primary cause of mismatches again.

Kimia-216 The third database has the highest significance to this project since its results can directly be compared to the ones originally proposed in [BL08]. The database by itself consists of 18 classes with respectively 12 instances. Table 8.4 shows some exemplary shapes taken from this set. According to the authors, the PSSGM had been performed excellently on this data set. Thus, any discrepancy between both outcomes provides important hints at latent obstacles which have not been considered during the implementation. The actual evaluation is realised as previous by taking the similarity ranked output that is going to be analysed in terms of precision and recall. Having the same configuration as before, the

Table 8.4: The table illustrates some object classes of Kimia-216 database as they have been already used in [BL08]. In analogy to this, the project recruits this database for the purpose of evaluating its re-implemented version of the PSSGM (cf. [Hed+13]).

algorithm does only achieve a precision of 0.818, a number that drastically differs from the performance presented in [BL08]. This becomes even more obvious by comparing the absolute numbers with those originally proposed (cf. Table 8.5). Besides this, the corresponding precision/recall curve, shown in Figure 8.2, illustrates the evolution of this precision in relation to the recall.

	1st	2nd	3rd	4th	5th	6th	7th	8th	9th	10th	11th
Original	216	216	215	216	213	210	210	207	205	191	177
Reconstructed	205	208	202	199	200	192	184	167	161	130	96

Table 8.5: The table shows the similarity ranked output that has been achieved by recruiting the re-implemented PSSGM (lower row). In comparison to this, the results of [BL08] are additionally depicted in the upper row of the table. A closer look at these arrangements reveals major differences between both versions towards their recognition performance.

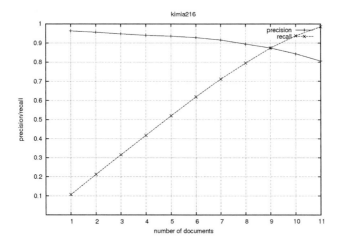

Figure 8.2: The figure illustrates the evolution of the precision relative to the recall that is successively increased towards the amount of target objects taken into account. The foundation of this analysis is given by the shapes of the kimia-216 database as originally used in [BL08] to evaluate the PSSGM.

Although the retrieval results are acceptable, the question for the reasons responsible for this divergence remains. The first intuitive assumption blames the skeletal structures which have been used during the evaluation, e.g. with regard to spurious branches. By checking all skeletons returned by the DCE, this expectation is confirmed. The review discovered plenty of shapes suffering from these violating outliers. Thus, it is no surprise that the retrieval result increased remarkably after manually pruning those branches which did not fit to the visual appearance of the corresponding shapes. Strictly speaking, this user-driven interaction pushed the average recognition performance by about 10.0 percent towards certain object classes, e.g. the birds enhanced their result from 0.69 to 0.81, while the camel shapes achieved an outcome of 0.73 in contrast to 0.63. However, such a post-processing is neither desirable nor applicable in practice but one more time stresses the profound impact of these structural anomalies on the matching.

Finally, it has not been possible to reconstruct the original result by the re-implementation of the PSSGM. However, please keep in mind that the actual configuration of the DCE was not reported in the paper and thus, it cannot be guaranteed that the same parameters have been employed in context of this project. Moreover, it is not known whether any modifications have been applied to the DCE code used in [BL08] compared to the one which is publicly available. Even though the PSSGM did not perform as expected, it results have been at least acceptable. In addition to this, the observations revealed one more time that the quality of the skeletal structure has a crucial influence on the overall recognition performance.

8.2 Chapter 04: 2D Object Retrieval Based on 2D Points and Curves

After getting in touch with 2D skeletons and their applications, this section presents the evaluation of a competitive approach using *contour points* and *contour curves* for the purpose of shape description. According to the information given in Chapter 4, configuration details of the evaluation framework are introduced first before the actual test results are addressed by the remaining content of this section.

8.2.1 Evaluation Setup

Despite the fact that almost all stages require user-defined inputs, the configuration of the evaluation environment is quite simple. Starting with the generation of partition points, the simplification of the contour polygon is fulfilled by exploiting the Douglas-Peucker algorithm with a tolerance value of: $\epsilon^{(DP)} = 4.0$. In contrast to this, the coefficients of the LPF are derived by a Gaussian distribution having a standard deviation of 1.0 which is going to be decreased incrementally by 0.1 in presence of false positives (cf. Chapter 4). In context of the feature generation, the Point Context is obtained by considering all contour points with the intent to populate a histogram with a dimension of 5×12 ($\log r \times \theta$). The curve-driven contour feature vectors are meanwhile managed without any additional input.

The final matching as well as the actual retrieval process takes place by employing $\alpha = 0.7$ as optimised weight to add the shape dissimilarities which have been returned independently from both feature types. Moreover, a method is employed that further improves the final retrieval result by analysing the underlying structure of the shapes' manifold. The benefit of this technique (published in [KDB10]) is its capability of capturing the data's manifold structure defined by the neighbourhood for each data point. Therefore, a modified version of a mutual kNN graph is recruited. Given a set of N shapes, the outcome of the N retrieval runs is gathered with the aim to create a $N \times N$ distance matrix carrying the pairwise relation of all shapes in terms of the underlying dissimilarity measure. Having this matrix, the approach exploits the fact that the distance to all other shapes already contains important information about the overall shape manifold. Finally, the evaluation results are compared to those of the PSSGM [BL08], its revised version [Hed+13] and to the work of [Yan+15b] which employs the same contour features. However, instead of using the partition points as elaborated in Section 4.2, the authors of [Yan+15b] utilise the end points

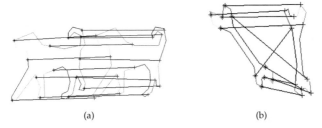

(a)   (b)

Figure 8.3: The figure shows two exemplary matching results obtained by the proposed method. The partition points are marked by the stars while the lines are used to indicate the final alignment between them. **(Left:)** A perfect matching result based on two instances of a dog. Please observe that even in presence of small deformations, all points have perfectly been aligned. **(Right:)** Contrary to this, a correspondence configuration with multiple false positives caused by the predominant symmetry between both bone shapes. All objects are taken from *Kimia99* data set.

of the skeleton to extract the corresponding contour fragments. In more detail, the DCE has been implemented together with a termination parameter of 4.0. This value was determined experimentally and leads to an equivalent number of partition points with regard to this technique.

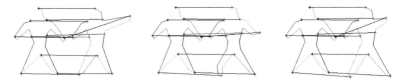

Figure 8.4: The figure shows multiple matching results established between three pairs of articulating shapes. While the query stays the same for all examples, different target instances have been selected. All object pairs are perfectly processed by the proposed technique.

8.2.2   Evaluation Results

In order to stay comparable to the methods stated above, only the first 10 matches are considered for each query element as introduced in [BL08]. Therefore, the result list is arranged in descending order according to the targets' shape similarities. These distance information, in turn, is derived from the matching results as explained in Chapter 4. Multiple exemplary alignments are illustrated in the Figures 8.3, 8.4 and 8.5. Please notice that the partition points are depicted by the stars, whereas the correspondence set is indicated by the lines between them. Although the final configuration of Figure 8.3b is corrupted by multiple misalignments, the algorithm performs quite robustly even in presence of deformations as illustrated in Figure 8.3a. The difference between these situations lies in the degree of discrimination that is carried by the shapes of both object classes. While the bone segments are mostly meaningless due to the overall symmetry, the boundary parts of the dogs are more distinguishable. Thus, the matching result is reasonable, although the configuration of Figure 8.3b seems to be entirely distorted. In addition to this, Figure 8.4 and 8.5 further emphasise the robustness of the proposed technique presented in Chapter 4. On the one hand, in presence of articulating object parts and one the other by exploiting a set of objects suffering form heavy deformations and not perfectly determined partition points.

Figure 8.5: The figure illustrates three different configurations of correspondences. The objects (Kimia-99) exhibit stronger deformations as well as not perfectly determined partition points. Even in presence of these obstacles, the method is capable of achieving good results.

Table 8.6: The table contains the retrieval results of the Kimia-216 data set. Besides the approach introduced in Chapter 4, four further state-of-the-art methods have been applied to these shapes. Although the proposed technique achieves excellent results on these object instances, it cannot exceed the final recognition performance of the PSSGM.

Object Retrieval for Kimia-216

Algorithm	1st	2nd	3rd	4th	5th	6th	7th	8th	9th	10th
PSSGM [BL08]	216	216	215	216	213	210	210	207	205	191
PSSGM [Hed+13]	205	208	202	199	200	192	184	167	161	130
CS [Yan+15b]	216	215	206	204	200	186	172	163	130	124
CS + Skeletons [Yan+15b]	216	216	214	213	213	211	204	193	184	175
DCE + CS [Yan+15b]	216	204	197	185	175	162	154	154	142	131
Proposed Method	**216**	**214**	**207**	**204**	**201**	**204**	**191**	**188**	**192**	**185**

Once the retrieval result is available, the targets (ordered according to their similarity) are checked in analogy to the other projects whether they belong to the class of the currently selected query (true positive) or not (false positive). By summing up all true positives in every column, the recognition performance can be analysed separately for each position inside the result list as depicted in Table 8.6 and 8.7. Both tables are devoted to the first ten matches of each retrieval result after applying the different methods respectively to the *Kimia-216* and the *MPEG-400* databases. Please also notice the abbreviations which have been used as method identifiers: Path Similarity Skeleton Graph Matching (PSSGM), Contour Segments (CS) and Discrete Curve Evolution (DCE).

Table 8.7: This table presents the second part of the evaluation operating on the shapes provided by the MEPG-400 database. Similar to Kimia-216, this data set has been processed by multiple state-of-the-art methods in order to achieve a more meaningful comparison. However, in contrast to the Kimia-216 data set, this time the proposed method did clearly outperform *all* of its competitors.

Object Retrieval for MPEG-400

Algorithm	1st	2nd	3rd	4th	5th	6th	7th	8th	9th	10th
PSSGM [Hed+13]	380	371	361	351	344	339	332	320	330	309
CS [Yan+15b]	375	348	333	325	317	311	300	295	276	275
CS + Skeletons [Yan+15b]	383	373	364	356	349	343	336	320	330	312
DCE + CS [Yan+15b]	375	368	346	337	338	323	308	297	286	276
Proposed Method	**389**	**374**	**368**	**368**	**358**	**347**	**344**	**339**	**346**	**330**

Taking a closer look at the these tables, it is obvious that the proposed technique is capable of achieving excellent results on both data sets. Even though the method could not exceed the retrieval performance of the PSSGM presented in [BL08], it clearly outperforms its re-implementation as well as all other competitors which have been applied to the Kimia-216 data set. In context of the MPEG-400 shapes, it actually returns the best retrieval result among all techniques. Nevertheless, the approach strongly relies on the quality of the partition points. Moreover, it requires attention in order to behave robustly, e.g. the generation of meaningful partition points, two shape descriptors, an optimised weight coefficient and the shape manifolds. Thus, depending on the underlying application, one should thoroughly decide if the (slightly) increased profit justifies the drop of the skeleton-driven approaches which *only* demand the extraction of a skeletal structure and a proper path description.

8.3   Chapter 05: 3D Object Retrieval Based on 3D Curves

Like the chapters before, this thesis part evaluates the method developed in Chapter 5. Strictly speaking, the section reviews the recognition performance as well as the robustness of the proposed method towards outliers and deformations. For this purpose, a 3D database of chairs, tables and stands was created. These object classes are highly similar to each other and thus, the database possesses the risk of ambiguities concerning the shape of the objects. In contrast to this, they fulfil the demands of the proposed method towards the requested geometrical properties, e.g. of being not fully opaque. The entire set of objects consists of 213 chair instances, 107 stands and 70 table objects. It is worth mentioning that the chairs are identical to those employed in [MYL13]. The same does also hold for the stands but the original set has additionally been extended by further instances. Using the same foundation of objects, the retrieval results become comparable to the outcome of [MYL13]. Please notice that all objects have been recorded with a RGB-D device from randomly chosen view points placed inside complex real-world scenarios. Three exemplary situations are illustrated in Figure 8.6. The Ground Truth (GT) for all of these objects has been generated manually and

Figure 8.6: The figure shows three exemplary scenes captured by a Kinect[©] sensor. Moreover, instances of all object classes are displayed which have been subjected in this project. Please recognise that some of the objects are occluded, arranged close to each other or only partially visible at the image borders.

labels each instance in accordance to the available classes: *chair, stand, table* and *undefined*. The actual object retrieval process is then performed by successively passing the three user-sketched queries to the system (cf. Figure 8.7). As already stated in Chapter 5, the structural composition of the queries are consciously kept simple in order to provide a convenient interface for the interaction with the end user. Figure 8.7 provides, in addition to the query objects, an impression of the measuring quality that can be expected during all tests.

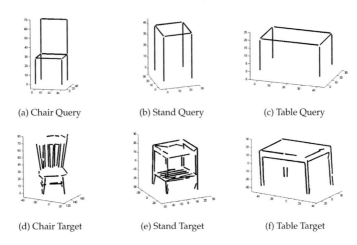

(a) Chair Query   (b) Stand Query   (c) Table Query

(d) Chair Target   (e) Stand Target   (f) Table Target

Figure 8.7: The figure provides an impression of the user-sketched queries and their corresponding real-world counterparts addressed by them. This is the chair, the stand and the table class. While the upper row exclusively illustrates the query instances, the lower one is dedicated to the measurements in their quality which can be expected during all tests.

8.3.1   Evaluation Setup

The next lines are discussing the system configuration that parametrises all values required by the proposed method. Although most of these numbers correlate to the *average* or the *standard deviation* of the underlying data, some of them are determined purely analytically. While the tolerances for the detection of feature points are an example for such an exclusively driven analysis (presented in the last paragraph of this chapter), the average length of all 3D curve fragments is incorporated for establishing virtual links between these features. In other words, let $\psi : \mathbb{R}^n \to \mathbb{R}$ be a function that accepts a sequence of numbers on which the average is generated, the following thresholds are determined by considering all curves during the creation of feature connections: $\mu^{(A)} = 0.4 \cdot \psi$, $\mu^{(R)} = 0.5$, $\mu^{(C)} = 0.3 \cdot \psi$ and $\mu^{(W)} = \pi/9$. The notion behind the incorporation of the average (or the standard deviation) is to stay scale-invariant inside the same object class. Even though the coefficients are estimated empirically, the problem of finding an appropriate configuration is additionally relaxed due to the modified Dijkstra algorithm that pushes the property of being insensitive to falsely detected virtual links. Moreover, it can be argued that the algorithm performs better in presence of additional spurious connections than in absence of those which are crucial for the overall shape appearance. Once all thresholds are generated, shortest paths can be localised with the intention of representing the objects' structures. Therefore, each path is scanned by 50 equally distributed sample points which are the basis for the preparation of descriptors. With access to these object characterising sequences, the system is fully prepared to start the evaluation. Here the DTW (and the EMD) is directly utilised for two tasks: (i) the calculation of the dissimilarity between two time series delivering an appropriate indicator for the *path distance* (cf. Section 5.5) and (ii) the estimation of the best fitting LCS as described in Section 5.3.1. Related to the point ordering scheme or rather the feature clustering approach, the standard deviation is exploited for identifying density regions of feature points (cf. Section 5.6.1). The final similarity values are then summarised by means of a precision/recall curve supporting a better assessment of the evaluation and, in addition to this, by the average precision (AP) (as defined in [Eve+10]).

8.3.2   Evaluation Results

Besides the required operation time, the actual recognition performance constitutes the most important number for the purpose of rating the success or failure of such a system. Commonly it is expressed in terms of *precision* and *recall* measured based on the result list. In parallel, this project does also calculate the AP to increase the comparability to other state-of-the-art approaches. After discussing the recognition performance for each data set respectively, this section concludes with a final benchmark investigating the capability of detecting feature points correctly.

Chair Database: As already mentioned at the beginning, the chair database is identical to that used in [MYL13]. Please be informed that this data set is highly challenging in view of the working principle behind the proposed method due to quality issues and outliers. Moreover, the set includes objects which cannot be recognised at all by the presented approach because of conceptual reasons, e.g. the shape of an office chair is not supported by the methodology of the feature detection method. However, even in the presence of this badly conditioned situation, the method is able to achieve excellent results with an average precision of **0.760** using the DTW to calculate the path distance. This is also demonstrated in Figure 8.8 that accommodates the resulting precision/recall graph and its corresponding ROC curve next to it. The figure clearly shows a perfect precision up to 40%, afterwards it quickly drops. Nevertheless, a recall of about 70% still returns a precision of approximately 50%. In addition to this, the ROC curve indicates a higher true positive rate attesting the procedure a strong discrimination power towards the currently selected object class. Finally, it has to be noted

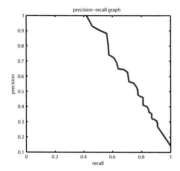 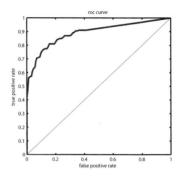

Figure 8.8: The figure illustrates the result of the proposed method after applying it to the chair database originally introduced in [MYL13]. The path distance calculation has been realised by DTW. **(Left:)** The corresponding precision/recall curve. **(Right:)** Receiver Operating Characteristic (ROC) curve derived from the same result displaying the relation of true to false positives.

that the proposed method even outperforms all of its state-of-the-art competitors. Table 8.8 provides an overview of all AP values which have been achieved by the different approaches.

Stand Database: After processing the chair instances, this paragraph is devoted to the stand assigned object class. The results of this evaluation step are presented again in analogy to the previous content. Please recall that the DTW has been selected to calculate the path distance values. The retrieval result is plotted in Figure 8.9 in form of a precision/recall graph as well as its ROC curve. At first glance, the result appears devastating and moderates the good impression which has been conveyed during the previous study. The reason for

Method		average precision
Proposed Approach		0.760
Ma et al.	[MYL13]	0.714
Janoch et al.	[Jan+11]	0.438
Felzenswalb et al.	[Fel+10]	0.419
Ferrari et al.	[FJS10]	0.351

Table 8.8: The table shows the AP value of the introduced method in comparison to other state-of-the-art techniques. The most relevant information is given by Ma et al. since their approach strongly relates to the proposed one.

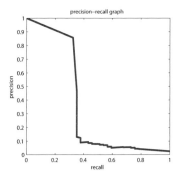
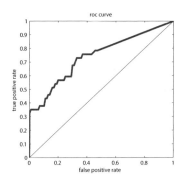

Figure 8.9: The figure illustrates the retrieval result of the stand query that has been received by the proposed method in combination with DTW. **(Left:)** The precision/recall graph that drastically drops at a recall of about 40%. **(Right:)** Its corresponding ROC curve.

this failure has to be ascribed to the design of the path descriptor (cf. Section 5.5). With the intent to stay scale invariant, the introduced path descriptors only encompass tuples of angle relations and thus, no information about structural proportions are stored. Consequently, the algorithm cannot discriminate the stand properly from the lower part of a chair. Perceiving this problem, the DTW has been replaced by the Earth Mover's Distance (EMD), as already suggested in Section 5.5, that additionally allows the incorporation of the vector lengths to each sample point. This further attribute is then also considered during the computation of the path distances. Strictly speaking, by changing the code as described above, the algorithm would lose its property of being scale invariant. Hence, all lengths are finally normalised with regard to the object's height. For more detailed information, please refer to Section 5.5. Figure 8.10 shows the result of this substitution. It is obvious that the performance could be improved drastically compared to the previous one on this data set. However, the actual result exceeds any expectations with an AP of 0.8484 since this adaptation cannot fully suppress all path ambiguities which still remain. Please further keep in mind that the stand evaluation has been performed on the extended database.

Table Database: This evaluation cycle is the only one that operates exclusively on a new database. Hence, it constitutes the smallest collection of shapes used in this project. However, the object class has been chosen deliberately. On the one hand, its geometry fulfils the required demands of the algorithm and on the other, it provides a similar shape to the stand and to the lower part of a chair. From a scientific point of view, this constellation has a more attractive character since it also monitors the behaviour of the method in such a challenging

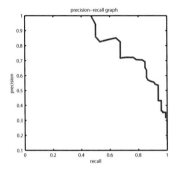 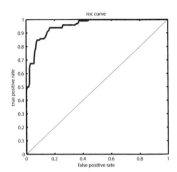

Figure 8.10: The figure presents the new retrieval result that has been obtained by substituting the DTW by the EMD. **(Left:)** The precision/recall graph showing a higher recognition performance. **(Right:)** The corresponding ROC curve. The AP is 0.8484.

situation. In addition to this, it allows the validation of the conceptual modification that has been implemented in context of the stand database. The recognition performance by itself is illustrated again using the same instruments as above (cf. Figure 8.11). Please be aware of the fact that the first evaluation run yields a similar response to that which has been observed for the stands when using the DTW. Consequently, the EMD did undertake the job of calculating the path distance values with the result of an AP of 0.8840. This is a quite acceptable outcome for this database and further votes for the application of the EMD.

Feature Detection Performance In order to assess the feature detection mechanism, this paragraph evaluates the robustness and the accuracy of the developed feature detection technique presented in Section 5.3. Therefore, a Ground Truth (GT) has been created manually by inspecting the entire chair data set. During this review, a certain amount of 3D curve segments has been labelled as potential candidates whose start or end points are expected locations towards specific features. Having access to this GT, the data is compared with the features returned by the automatically operating routine responsible for their identification. If the location of such a point is close to those being part of the GT, this match is marked as *valid*. The result of this assessment is presented in Figure 8.12 which coincides with the observations gathered during the previous evaluations performed on the different object data sets. The excellent outcome is even more impressive when the poor quality of the input data is taken into account. The actual evaluation considers multiple combinations of parameters which have been tested. Please keep in mind that the symbol $\mu 1$ controls the triangular condition, whereas $\mu 2$ is managing the rectangular fitting procedure. While most

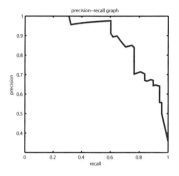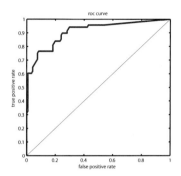

Figure 8.11: The figure shows the final outcome that has been monitored for the task of recognising 3D tables. The AP is 0.8840 and has been accomplished by replacing the DTW with the EMD. **(Left:)** The precision/recall graph showing an acceptable retrieval result. **(Right:)** Plot of the corresponding ROC curve.

of the configurations behave quite similar to each other and yield a magnificent detection performance in terms of accuracy and completeness, only the set of $\mu 1 = 0.3$ and $\mu 2 = 0.1$ seems to be too restrictive in its rectangular condition.

8.4   Chapter 06: 3D Abdominal Aorta Registration Based on 3D Skeletons

This section continues the discussion about the processing pipeline introduced in Chapter 6. In detail the focus is on the evaluation of the segmentation, skeletonisation and the actual matching. The data that is exploited for this purpose has been provided by the SOVAmed GmbH and consists of nine pairs of CTA scans showing the abdominal aorta of real patients.

8.4.1   Evaluation Setup

In the following, the setup of the evaluation is addressed by introducing all values which are required to perform the registration of the vascular structures. In contrast to the other projects, this part is a bit more sophisticated and requires more attention. Please keep in mind that the 18 CTA series are nine pairs of pre- and post scans having a resolution of (512×512) per slice. The volume (the number of slices) varies in a range of 220 to 680. Moreover, it is not guaranteed that these attributes are identical even for pairs of corresponding scans.

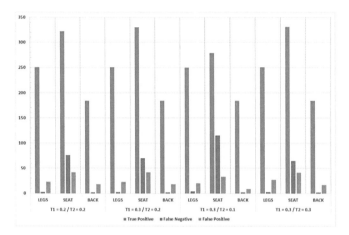

Figure 8.12: The figure provides an overview of the detection performance towards different configurations. In total, four different combinations of thresholds are evaluated with the aim to assess the detection capability in terms of accuracy and completeness. The diagram shows the reader the amount of detected features by counting the *true positives*, the *false negatives* and the *false positives*. Altogether, all pairs are performing quite well, except of the one with $\mu1 = 0.3$ and $\mu2 = 0.1$. Here it seems that the values are too restrictive.

Segmentation   In order to increase the precision of the segmentation step, an anisotropic diffusion filtering is applied to the CTA series in advance. Therefore, a conduction coefficient function, proposed by Perona et al. in [PSM94], is employed:

$$\frac{1}{1+\left(\frac{\|\nabla I\|}{\nu}\right)^2} \quad , \tag{8.1}$$

with ν being the gradient modulus threshold controlling this conduction process. The value of it is experimentally determined and finally set to 70. Afterwards, the segmentation approach is applied to this normalised data. In context of the surface reconstruction, the initial number of clusters is set to 5. Moreover, a boundary feature map (related to the image gradient) is required for the execution of the hybrid level set segmentation technique [Zha+08]. This map is defined as:

$$g = \frac{1}{1-a\|\nabla I\|^2} \quad , \tag{8.2}$$

where the constant a is set to 5 controlling the slope of the function. The other parameters in Equation (6.24) are configured with $\eta = 0.5$ and $\tau = 0.2$, respectively. This setting allows to efficiently segment the aorta in all 18 series. An exemplary segmentation result is shown in Figure 8.13.

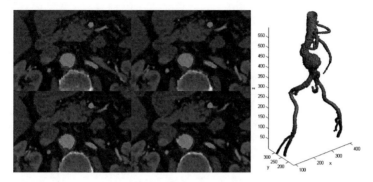

Figure 8.13: The figure illustrates the working principle of the segmentation approach as well as an exemplary instance of a fully segmented abdominal aorta. **(Left:)** Four consecutive slices showing three clusters (drawn in green) which are traced by the method. **(Right:)** The fully segmented vascular structure in 3D.

Skeletonisation   The implementation of the skeletonisation method introduced in Section 6.3 uses the publicly available Matlab© source code[2] that mainly coincides with the explanations given in [VUB07]. However, it also consists of some improvements at certain places which are addressed in the following. Moreover, small adjustments were required by the input data in order to allow the extraction of the centre line from the vascular structure. In summary, the speed image is generated in accordance to Equation (6.36) and the termination criteria of the algorithm does also follow the recommendation of the authors. That is, the maximum value of the time-crossing map T is taken twice with the intent to approximate the maximum diameter of the vessel. In contrast to this, the interpolation scheme is inherit from the improvements of the available code using the *Runge-Kutta* method[3]. However, the pre-defined step size of 1.0 had to be changed due to the reasons discussed in Section 6.3. Beyond that, it is worth knowing that the fast marching approach is also borrowed from

[2]http://www.mathworks.com/matlabcentral/fileexchange/24531-accurate-fast-marching, [online: 19th August 2015]
[3]http://mathworld.wolfram.com/Runge-KuttaMethod.html, [online: 19th August 2015]

the code and utilises the technique presented in [HF07a] to enhance the accuracy of the output. An exemplary outcome of this hybrid version is demonstrated in Figure 8.14 where the centre line clearly reflects the structure of the original aorta.

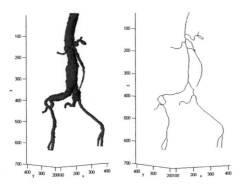

Figure 8.14: The figure shows an exemplary skeleton returned by the skeletonisation approach as it is used in this project. **(Left:)** Vascular structure generated as output of the segmentation approach. **(Right:)** Skeleton which has been extracted from the (segmented) input given on the left.

Matching   While the Hungarian method did not require any parameters, the MWC method is configured as follows: (i) The Gaussian function is parametrised by (0.5; 5.0). Please be aware that the first value is used as scaling coefficient (the curve's peak), whereas the second denotes its standard deviation. (ii) The value of θ (binary potentials) is set to 0.5 and (iii) the threshold for establishing mutex constraints is set to 15 voxels in each dimension. Moreover, 50 supports are employed to sample each of the shortest paths. This number was proven to deliver a sufficient amount of meaningful information during the last projects.

8.4.2   Evaluation Result

All pairs of abdominal aortas have been registered based on the Hungarian method and the MWC approach. In order to evaluate the matching performance of these methods, these results have been verified by an expert qualified to check the correspondences towards their anatomical correctness. Therefore, the corresponding skeleton points on both series are marked, so that the expert was able to assess them visually. Figure 8.15 demonstrates such a labelled version of an arbitrary matching result.

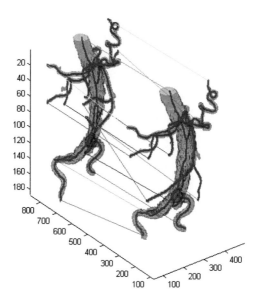

Figure 8.15: The figure illustrates an arbitrary matching result whose alignment configuration has been labelled visually by coloured lines in order to support the expert during his assessment of the matching result. Moreover, the segmented vascular structure is displayed to provide a better visual feedback.

Depending on the CTA series, all analysed data sets led to 7 to 14 skeleton end points that had to be matched during the registration. With access to these points, the feature generation starts to generate descriptors for all shortest paths established with the intent to characterise the skeletal structure. Once these descriptors are available, the actual evaluation of the matching is performed based on two cycles. In the first run, the registration is realised by the Hungarian method which is subsequently replaced by the MWC approach (second cycle). While the analysis of the Hungarian method discovers a result of 6 completely correctly assigned instances having an average accuracy of 87.84% properly matched end points (65 of 74), the MWC approach accomplishes a slightly better result. Instead of 6, the MWC technique is able to register 7 of 9 series without any mismatches and reaches a mean accuracy of 91.9% true positives (68 of 74). False positives and/or false negatives are only monitored in situations where the resolution between both series differ from each other or where the vessels are of different lengths. It is worth mentioning that the MWC approach hosts the potential risk that its binary relations are working contrary to its mutex constraints and vice versa. Strictly speaking, correlating with the power which is either given to the one or to the other condition, it might happen that true positives are forced to be misaligned while false positives are changing their state to true positives at the same time.

8.5   Chapter 07: 3D Object Retrieval Based on 3D Curve Skeletons

This thesis part presents the results which have been obtained by the method introduced in Chapter 7. Since the focus of this project was placed on the generation of curve skeletons, one point of this evaluation concerns the rating of these structures. Therefore, Ground Truth (GT) has been generated by human volunteers which classified the similarity of objects being part of the underlying database. Please notice that the human cognition is not binary and that humans are able to rate the magnitude of similarity by fuzzy constraints based on their world knowledge, a fact that had to be considered during the design of the GT Section 8.5.1. In addition to this, the performance of the retrieval system is once more rated based on the indicators *precision* (accurateness) and *recall* (completeness) which, of course, requires the existence of the GT.

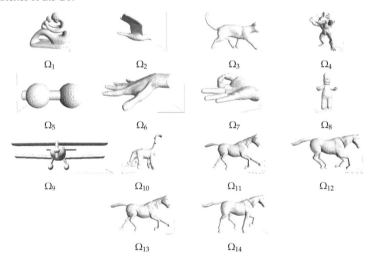

Figure 8.16: The figure gives an overview of the objects used for evaluating the system performance. The entire data set consists of 14 instances where the objects' geometries mostly consist of articulating parts.

8.5.1 Evaluation Setup

As already stated above, the Ground Truth in this section is generated by human volunteers instructed to rate the similarity between 91 combinations which are generated based on 14 objects (cf. Figure 8.16). However, the human interaction does discover several issues: First, human beings are used to classify their impressions according to the states *similar* or *not similar*, even though there exists a fuzzy logic behind this process. One has to be aware of this if people are asked to express the degree of vagueness in their decisions in form of a deterministic number. Second, each person is influenced by the *word knowledge* which they (mostly) gathered in their past. That means, object pairs might be rated differently by different persons due to latent connections which are part of this knowledge. However, in order to stay comparable to the cosine-driven output delivered by the proposed system, the volunteers are briefed as follows: (i) All similarity values have to be placed in a range of $[0 - 1]$, (ii) identical objects have to be rated with 1.0 and (iii) geometrical properties like, e.g. the number and the proportion of shape segments shall be taken into account during this assessment. The actual expert group encompasses 15 students coming from different disciplines. This heterogeneous composition follows the aim to obtain a statistically independent distribution of ratings. Table 8.9 provides the averaged similarity values for each pair of objects inside the database. Please notice that the coloured numbers are the most interesting ones since they represent those object pairs which are either considered as highly similar (red) or classified as almost dissimilar (blue) by the expert group.

Table 8.9: The table provides the averaged similarity values for each pair of objects. These numbers have been obtained by an expert group consisting of 15 human volunteers.

	Ω_1	Ω_2	Ω_3	Ω_4	Ω_5	Ω_6	Ω_7	Ω_8	Ω_9	Ω_{10}	Ω_{11}	Ω_{12}	Ω_{13}	Ω_{14}
Ω_1	-													
Ω_2	0.10	-												
Ω_3	0.12	0.20	-											
Ω_4	0.20	0.07	0.27	-										
Ω_5	0.12	0.20	0.13	0.05	-									
Ω_6	0.13	0.12	0.15	0.13	0.17	-								
Ω_7	0.13	0.11	0.12	0.10	0.17	0.90	-							
Ω_8	0.19	0.07	0.40	0.47	0.10	0.20	0.13	-						
Ω_9	0.08	0.46	0.07	0.07	0.09	0.07	0.10	0.06	-					
Ω_{10}	0.22	0.07	0.38	0.30	0.07	0.10	0.17	0.17	0.07	-				
Ω_{11}	0.13	0.10	0.61	0.23	0.07	0.07	0.10	0.19	0.07	0.60	-			
Ω_{12}	0.13	0.10	0.60	0.23	0.10	0.07	0.10	0.19	0.07	0.59	0.95	-		
Ω_{13}	0.12	0.10	0.63	0.23	0.10	0.07	0.10	0.19	0.07	0.59	0.95	0.92	-	
Ω_{14}	0.13	0.10	0.60	0.23	0.10	0.07	0.10	0.19	0.07	0.61	0.96	0.96	0.95	-

Closely inspecting Table 8.9 confirms a rather inconspicuous result. The highest similarity values are monitored for the group of the horses as well as for the hand models.

Both observations are reasonable observations which can easily be justified by the shape conformity among these objects. More interesting however, is the result of Ω_5 to Ω_6 and Ω_7. Although a similarity of 0.17 is still rather low in fact, it should not be neglected since its magnitude is quite surprising anyhow for two geometry models (dumbbell/hand) which are entirely different to each other. An explanation for this blames the latent world-knowledge of each human being connecting the action of lifting a dumbbell with the model of a hand. Even though it is obvious that the method is not going to rate these objects as similar, the information is kept during the actual retrieval since it reflects the cognitive skill. Finally, the data in Table 8.9 is utilised to generate the desired GT. Therefore, a heuristically operating procedure is proposed capable of forming a subset of objects that is then considered as true positives for a currently selected one. Moreover, it is worth knowing that this GT is varying in its size because the database is not balanced in its the number of instances towards every object class. The actual estimation is then driven by the differences of similarities:

Step 1:   For each object Ω_h all remaining instances are arranged in a descending order according to their similarity (cf. Table 8.9) with regard to Ω_h. The result of this operation, namely an ordered sequence of tuples, is depicted by the symbol $\hat{a}$ in the following: $\hat{a} = \{(\Omega_{i,0}, a_{i,0}), \ldots, (\Omega_{i,j}, a_{i,j}), \ldots, (\Omega_{i,12}, a_{i,12})\}$, with $a_{i,j-1} > a_{i,j}$.

Step 2:   Afterwards, difference values are calculated between adjacent members in $\hat{a}$ by traversing all similarity entries successively: $\psi(a_{i,j-1}, a_{i,j}) = a_{i,j-1} - a_{i,j}$.

Step 3:   By localising that difference value with the fourth largest magnitude and back-tracking its position to $\hat{a}$, a (spatial) border is detected masking all objects above as members of the GT (cf. Table 8.10). Strictly speaking, let $\mathcal{K}_h$ be the set of GT elements towards the query Ω_h and further assume that the searched difference is generated by $\psi(a_{i,k-1}, a_{i,k})$ with $k = [0, \ldots, (\|\hat{a}\| - 2)]$, then $\mathcal{K}_h = \{\Omega_i \mid (\Omega_{i,j}, a_{i,j}) \in \hat{a} : j > k\}$.

8.5.2   Evaluation Results

After creating an appropriate GT, this section evaluates the performance of the system. The objects used in this context are shown in Figure 8.16. It is obvious that this collection constitutes a quite challenging but suitable set for investigating the discrimination power of the topological oriented descriptors (cf. Section 7.3) as well as the skeletonisation process as introduced in Section 7.2. In more detail, the objects' geometries mostly consist of articulating parts and naturally tend to produce surface skeletons.

Table 8.10: The table demonstrates the heuristical selection of objects responsible for generating the Ground Truth $\mathcal{K}_h$ of a certain object Ω_h based on a constructed set of similarities.

list	$\hat{a}^{\top}$	$\psi(a_i, a_{i-1})$		$\mathcal{K}_h$
$\Omega_{i,0}$	0.9	-	-	√
$\Omega_{i,1}$	0.65	0.25	1	√
$\Omega_{i,2}$	0.64	0.01	7	√
$\Omega_{i,3}$	0.45	0.2	3	√
$\Omega_{i,4}$	0.3	0.15	4	x
$\Omega_{i,5}$	0.06	0.24	2	x
$\Omega_{i,6}$	0.02	0.04	5	x
$\Omega_{i,7}$	0.0	0.02	6	x

Taking these objects, the aim is to determine the precision and the recall by applying the proposed method to this set. Therefore, similarity values are first calculated by extracting the curve skeleton for each object, generating its corresponding feature vector and passing it together with the one of the target to the cosine measure (cf. Section 7.1.2). The outcome of this procedure is presented in Table 8.11. With access to these numbers, a result list for each object is derived by exploiting the same heuristic approach as already employed in context of the ground truth (cf. Section 8.5.1). Having both the result list and the corresponding GT, the recall and the precision indicators are computed for all objects as illustrated in Table 8.12. Altogether, the retrieval outcome is rather moderate and is mainly suffering from

Table 8.11: The table provides an overview of the similarity values which have been returned by the system based on the proposed similarity measure fed by the feature vectors introduced in Section 7.3.

	Ω_1	Ω_2	Ω_3	Ω_4	Ω_5	Ω_6	Ω_7	Ω_8	Ω_9	Ω_{10}	Ω_{11}	Ω_{12}	Ω_{13}	Ω_{14}
Ω_1	-													
Ω_2	0.704	-												
Ω_3	0.966	0.789	-											
Ω_4	0.981	0.682	0.946	-										
Ω_5	0.476	0.699	0.401	0.511	-									
Ω_6	0.928	0.811	0.991	0.896	0.349	-								
Ω_7	0.944	0.798	0.992	0.900	0.351	0.994	-							
Ω_8	0.980	0.644	0.922	0.995	0.522	0.863	0.876	-						
Ω_9	0.987	0.693	0.954	0.997	0.506	0.912	0.914	0.898	-					
Ω_{10}	0.982	0.670	0.939	0.999	0.515	0.887	0.894	0.998	0.995	-				
Ω_{11}	0.988	0.749	0.989	0.980	0.450	0.961	0.967	0.968	0.982	0.978	-			
Ω_{12}	0.989	0.683	0.952	0.994	0.496	0.902	0.915	0.995	0.989	0.996	0.986	-		
Ω_{13}	0.986	0.679	0.948	0.995	0.499	0.897	0.908	0.996	0.989	0.997	0.984	0.999	-	
Ω_{14}	0.992	0.710	0.969	0.991	0.479	0.923	0.938	0.988	0.989	0.991	0.995	0.999	0.997	-

conceptual problems: First, the database used during this evaluation is highly unbalanced and does not always provide a second instance for each object class. Thus, the algorithm is not able to find a *real* counterpart for every query. Second, the heuristic approach does not perform well in all cases, e.g. the GT collection of a horse does only encompass one instance despite the fact that two further ones are available. However, related to the human cognition, the method delivers quite promising results in view of rating similarities between 3D objects and provides the potential to improve the overall system performance. Apart from this, two additional points are observed which have to receive further attention: (i) the group of horses consists of multiple outliers leading to an unexpected and rather poor precision. Strictly speaking, even though all horses were retrieved by the algorithm, there are multiple instances of wrong classified objects. The problem has to be ascribed to the proposed feature vector that only processes the topology of the skeleton. Consequently, the power of discrimination is reduced towards those instances which only differ in their articulation. The second issue reflects the problem of world-knowledge characterised by latent connections of conceptual relations between objects, e.g. the hand and the dumbbell. As already expected, these objects are not considered as similar by the method.

Table 8.12: The table shows the recall and the precision values obtained by the proposed method. It is obvious that the system performance is rather moderate.

Query	True Positives (GT)	System Classified	Recall	Precision
Ω_1	$\Omega_2, \Omega_3, \Omega_4, \Omega_5, \Omega_6, \Omega_8, \Omega_{10}, \Omega_{11}, \Omega_{12}, \Omega_{13}, \Omega_{14}$	$\Omega_3, \Omega_4, \Omega_8, \Omega_9, \Omega_{10}, \Omega_{11}, \Omega_{12}, \Omega_{13}, \Omega_{14}$	0.73	0.89
Ω_2	$\Omega_3, \Omega_5, \Omega_6, \Omega_7, \Omega_9$	$\Omega_1, \Omega_3, \Omega_4, \Omega_5, \Omega_6, \Omega_7, \Omega_9, \Omega_{10}, \Omega_{11}, \Omega_{12}, \Omega_{13}, \Omega_{14}$	1.00	0.42
Ω_3	Ω_{11}, Ω_{13}	$\Omega_7, \Omega_9, \Omega_{11}$	0.50	0.33
Ω_4	$\Omega_1, \Omega_3, \Omega_8, \Omega_{10}, \Omega_{11}$	$\Omega_1, \Omega_8, \Omega_9, \Omega_{10}, \Omega_{11}, \Omega_{12}, \Omega_{13}, \Omega_{14}$	0.80	0.50
Ω_5	Ω_6, Ω_7	$\Omega_1, \Omega_2, \Omega_3, \Omega_4, \Omega_7, \Omega_8, \Omega_9, \Omega_{10}, \Omega_{11}, \Omega_{12}, \Omega_{13}, \Omega_{14}$	0.50	0.08
Ω_6	$\Omega_1, \Omega_2, \Omega_3, \Omega_5, \Omega_7, \Omega_8, \Omega_{10}$	$\Omega_3, \Omega_7, \Omega_{11}$	0.29	0.67
Ω_7	$\Omega_1, \Omega_2, \Omega_5, \Omega_6, \Omega_8, \Omega_9, \Omega_{10}$	Ω_3, Ω_6	0.14	0.50
Ω_8	$\Omega_1, \Omega_3, \Omega_4, \Omega_6, \Omega_{10}, \Omega_{11}$	$\Omega_1, \Omega_4, \Omega_9, \Omega_{10}, \Omega_{11}, \Omega_{12}, \Omega_{13}, \Omega_{14}$	0.67	0.50
Ω_9	$\Omega_2, \Omega_5, \Omega_7, \Omega_8, \Omega_{11}$	$\Omega_1, \Omega_4, \Omega_8, \Omega_{10}, \Omega_{11}, \Omega_{12}, \Omega_{13}, \Omega_{14}$	0.40	0.25
Ω_{10}	Ω_{11}	$\Omega_1, \Omega_4, \Omega_8, \Omega_9, \Omega_{11}, \Omega_{12}, \Omega_{13}, \Omega_{14}$	1.00	0.13
Ω_{11}	Ω_{13}	$\Omega_1, \Omega_3, \Omega_4, \Omega_9, \Omega_{11}, \Omega_{12}, \Omega_{13}, \Omega_{14}$	1.00	0.13
Ω_{12}	Ω_{14}	$\Omega_1, \Omega_4, \Omega_8, \Omega_9, \Omega_{10}, \Omega_{11}, \Omega_{13}, \Omega_{14}$	1.00	0.13
Ω_{13}	Ω_{11}	$\Omega_1, \Omega_4, \Omega_8, \Omega_9, \Omega_{10}, \Omega_{11}, \Omega_{12}, \Omega_{14}$	1.00	0.13
Ω_{14}	Ω_{12}	$\Omega_1, \Omega_3, \Omega_4, \Omega_8, \Omega_9, \Omega_{10}, \Omega_{11}, \Omega_{12}, \Omega_{13}$	1.00	0.11

Chapter 9

Summary, Conclusion and Outlook

The last section of this thesis provides a comprehensive summary and collects the most interesting findings which could be obtained during the development and evaluation of the single projects being part of this work. Moreover, an outlook is given on future tasks which are either in a pending state or only of theoretical nature.

9.1 Summary

Researchers realised early that incorporating 3D information into the area of object detection and recognition has the potential to simplify and to improve these tasks in contrast to only using 2D. Besides an increased performance in terms of robustness and accuracy, algorithms can be designed to be view invariant and more insensitive to changing light conditions. Inspired by the work of Zygmunt Pizlo and by the research of Longin J. Latecki, the underlying thesis is primarily devoted to the use of 3D information coupled with a sophisticated shape descriptor, namely the skeleton.

Excellent results, demonstrated in [BL08], show that skeletons are an appropriate instrument to realise the task of object categorisation in 2D. This assumption has been confirmed on top of a novel approach, the Path Similarity Skeleton Graph Matching (PSSGM), proposed by the authors Bai and Latecki. The idea of the PSSGM is quite elaborated: Using the skeletal structure of an object, its shape complexity is reduced from 2D to 1D. On the one hand, the skeleton branches are emphasising the significance of boundary parts which are important for the perceptual appearance, on the other hand, it simplifies the feature generation process. The latter is based on the concept of shortest paths which are established in both skeleton graphs: $G^{(query)}$ and $G^{(target)}$. By exploiting an intuitive sampling procedure, each path is represented by a finite number of scalar values and thus, the entire skeletal structure. In detail, for each sample point a maximum disk is calculated that entirely fits into the boundary of the object at this position. The actual matching costs for a certain pair (v, u) (with $v \in G^{(\text{query})}$ and $u \in G^{(\text{target})}$) is then calculated by considering the dissimilarity between all emanating paths which respectively start at v and u. Finally, the optimal matching is found by the Hungarian method. Once this method had been studied in theory, the first project of this work was instructed with a deeper analysis of the PSSGM. By re-implementing the algorithm, the first practical contact could be established. In addition to this, the opportunity to figure out its strengths and weaknesses has been used. Although the reconstructed code is not capable of reproducing the same excellent results as presented in [BL08], the evaluation still provides a good recognition performance. Besides this, three critical points were observed during a thoroughly performed evaluation. Being aware of these issues further activities based on this matching approach were studied. Strictly speaking, the algorithm leads to serious problems in context of *flipped images, spurious skeleton branches* and the enforcement of *1-to-1 matchings*.

After encountering skeletons, further investigations concerning the skeletal structure were undertaken by comparing its performance towards other shape descriptors, namely *contour points* and *contour curves*. Therefore, a sophisticated technique was developed for the task of shape analysis returning a set of points which indicate those contour parts having the highest contribution to the object's visual appearance. Taking this set of points, the corresponding contour segments can easily be derived by slicing the boundary at these locations into multiple sub-segments. In order to encode the information of these primitives, two different feature sets are employed. One of them is an adaptation of the popular Shape Context, while the other recruits a ten-dimensional feature vector taken from [Yan+15b] that is capable of representing a certain contour segment based on geometrical ratio numbers. Afterwards, the shape dissimilarity is calculated independently for each descriptor by the Hungarian method. The proposed method has only a slightly worse recognition performance than the presented one in context of the PSSGM. The approach performs even better than the re-implementation of the PSSGM as introduced in [Hed+13]. Nevertheless, it is worth considering that the PSSGM only requires the extraction of the skeletal structure and an adequate sampling scheme for its branches. In contrast to this, the presented shape-driven technique demands more attention in order to behave robustly.

The first application which utilised the concept of the PSSGM method in 3D was designed in the light of detecting 3D chairs, stands and tables. Therefore, a natural scene is acquired by a RGB-D device. Afterwards, the resulting depth image is used together with a 2D edge detector aiming at the generation of 3D curve segments which are first linked and then back-projected into 3D space. In order to apply the idea of the PSSGM now, these disjoint 3D line fragments are interpreted as a special kind of skeletal structure. By further introducing a local coordinate system, "skeleton" end points are localised and subsequently connected based on virtual links established in a geometry-guided manner. Having this network of virtual links, the concept of shortest paths can be applied to this information in order to perform the actual matching realised by replacing the Hungarian method with a new matching technique, namely the Maximum Weight Cliques approach. The second project with a practical and even more important background arose from the field of professional medical diagnosis. It proposes a sophisticated processing pipeline to register two CTA series of the abdominal aorta. While the first scan displays the aorta beforehand where it is still affected by an aortic aneurysm, the second one shows the same vessel tree after the medical intervention. By aligning and/or overlaying the vessels of both scans, physicians shall be supported during the after-care of their patients. Therefore, the method starts with the segmentation of the vascular structure for the purpose of retrieving its centre line. The extraction of the skeleton is mandatory in order to apply the working principle of the PSSGM to this project. With access to this skeletal structure, two different feature sets are generated, the *intrinsic* and the *extrinsic* one. Finally, the matching is performed in a competitive setup evaluating both the Hungarian and the MWC approach.

After collecting a variety of experiences during the development and the evaluation of the projects in this thesis, two points can be emphasised: (i) Curve skeletons are an appropriate instrument for representing 3D objects while they are additionally reducing the complexity of the object's geometry. (ii) The extraction of curve skeletons is sophisticated and it depends on the object's class or rather geometry whether they can be extracted natively or not. Moreover, it is obvious that the PSSGM approach strongly relies on both the existence and the quality of the skeleton. In order to underline the importance of this condition, the last project is devoted to the evaluation of a further skeletonisation method capable of producing curve skeletons from a wide range of different, *not* tubular shaped 3D objects. Its working principle exploits so-called Jordan curves in order to localise object points which are forming a curve skeleton. Using this approach, excellent results were achieved on a set of objects whose geometries typically tend to surface skeletons. Instead of employing the PSSGM, a topological feature set was investigated coupled with a ground truth that was labelled by a group of human participants. Besides the investigations towards the generation of curve skeletons in general, synergies were searched which argue for the combination of topological and geometrical descriptors.

9.2 Conclusion

Taking into account that promising results could be achieved during the experiments which were the subject of this thesis coupled with the corresponding observations and findings, this work has successfully gathered a sufficient amount of data to be capable of replying appropriately to the questions stated preliminarily. In the following, these questions are going to be picked up again for the purpose of answering them in detail.

"Are (curve) skeletons *generic* enough concerning the task of 3D object *representation?*".

Yes, they are. Two different types of skeletons are available to accomplish the task of 3D object representation, namely *surface skeletons* and *curve skeletons*. While the concept of surface skeletons more strongly coincides with the definition of 2D skeletons, the curve skeletons provide a higher affinity to the human intuition and to the visual appearance of their 2D counterparts. Even though the thesis is devoted to the latter, the extraction of both structures is complex and difficult task. Additionally, the actual skeletonisation process is impeded by discretisation artefacts and noise causing, e.g. spurious skeleton branches. In addition to this, the generation of a curve skeleton requires more specialised approaches in presence of non-tubular geometry parts since these areas naturally tend to collapse to a surface during their skeletonisation. Nevertheless, the research of Dennie Reniers [Ren09] as well as own investigations showed that curve skeletons can be obtained from a wide variety of 3D objects. Moreover, it has been demonstrated how to map skeleton specific properties to other representation types to employ them in skeleton related applications.

"Are (curve) skeletons *meaningful* enough concerning the task of 3D object *matching* and *recognition?*".

It depends. Like other skeletal representation types, curve skeletons count to the family of shape descriptors. In its pure form, which is a composition of skeleton branches, the structure cannot be natively processed by almost every matching method. However, by considering the skeleton as an intermediate layer, it stresses those boundary parts with an enlarged significance to the visual appearance. Hence, it provides the perfect foundation for the feature generation task whose outcome is subsequently accepted by plenty of matching techniques. In addition to this, the transformation of the skeleton properties to a proper set of distinctive values does also lead to the description of the overall shape. Please notice that the actual descriptor design can strongly differ in terms of complexity and meaning as demonstrated in this work. Moreover, the emphasis of object characterising boundary regions allows a better performance in presence of deformations which might appear in articulating object parts. Nevertheless, the quality of the skeletal structure is crucial no matter which feature type is used. Structural anomalies, e.g. spurious skeleton branches or unintended loops, confuse the matching procedure with a drastic impact on the final result. Another quality related issue is presented in [VUB07], where the skeleton is determined on subvoxel precision in order to improve the alignment of the centre line.

"What are the *prospects* and *limitations* of such a concept intending to solve the task of 3D object matching?".

As described above, there is a huge potential towards the combination of 3D data with skeletal structures and graph based matching approaches that even allows the implementation of sophisticated object categorisation tasks. Although the instance of an object cannot be determined at this point due to the shape-driven character of the skeleton, it is worth mentioning that the skeletal structure provides the capability of being enriched by further object specific properties. This information would then be available during a possible object instance recognition task. This degree of flexibility enables the concept to be applied in a variety of working and research areas. Concurrently, one should not ignore difficulties which might additionally occur in these new situations. The skeletons' sensitivity to noise requires robust and smart operating strategies concerning the skeletonisation. Moreover, a concept has to be elaborated on how to reconstruct or to acquire the entire shape of an object for the purpose of extracting its skeleton. Especially this point might be critical in case of camera systems designed in a stationary fashion. Solving this problem does also constitute one of the future tasks in the next section. The idea is to establish an iteratively working skeleton construction process based on multiple (overlapping) views. In addition to this and depending on the application, it will be challenging to find an appropriate set of features that satisfies the necessity of being discriminative and adaptable at the same time. Closely inspecting the amount of computation time that is currently needed to perform a test run discovers an already increased demand of processing capacities. Nevertheless, by considering all these restrictions and by adjusting the right parameters, this concept allows the implementation of a well operating and powerful recognition system.

9.3   Future Work

In the following, a list of future tasks is given. Some of them are already in a pending state while others are only the derivative of theoretical deliberations. However, the most general and obvious task is the extension of all object databases which have already been used in this work. Moreover, it has to be guaranteed that further data sets of upcoming projects encompass an adequate amount of instances. Having access to such well-sized databases is important for realising a quantitative and qualitative evaluation.

Incorporation of Further (Perceptional) Properties Besides the depth data, further object properties can be incorporated into the representation. By developing a model capable of carrying detailed information about the object like colour, texture, motion, speed or even multispectral data, the algorithm is enabled to automatically select or weigh the most discriminative ones.

Object Instance Recognition Taking this over-representation, first investigations can be performed to evaluate the performance towards object instance-level recognition tasks. Connected to this, it can be evaluated whether all branches or only certain ones of a specific skeleton are needed to be enriched or even to be considered at all.

Surface Skeletons Instead of using curve skeletons exclusively, further research activities have to be conducted based on surface skeletons. Moreover, it has to be investigated to what extend the Path Similarity Skeleton Graph Matching can be applied to this kind of structure. Strictly speaking, the notion of skeleton end points as well as the concept of shortest paths have to be mapped on this surface-dominated shape in order to exploit the working principle known from the PSSGM.

Iterative Skeleton Refinement Being aware of the problem that a proper skeleton can only be extracted from an almost complete object model, further efforts have to be made on solving this problem. In cases which exclusively allow the acquisition of objects from a single perspective, the objects' boundaries are not fully accessible. To tackle this issue, an iteratively operating strategy has to be invented capable of constructing and refining the structure of a skeleton (and other specific object properties) over time. One solution might be the gathering and registration of multiple object perspectives until an adequate amount of data is available to reconstruct the skeletal structure like proposed in [TZCO09].

Partial Skeleton Matching This work package assumes that the 3D object model is not completely available but significant parts of its shape and its properties. Thus, it is worth evaluating the prospects and limitation towards the following points: (i) Is it possible to extract a kind of *sub-skeleton* from the data currently available by smartly bridging the gaps inside this incomplete scan? (ii) Are these skeletal structures sufficient to establish a partial matching with either an already trained skeleton representing the complete object or other partial views of the same object?

Please notice that this list only constitutes an outline of the most interesting points which have been observed during the development and evaluation of the projects being part of this work. The actual research topic devoted to the accomplishment of sophisticated recognition tasks using the synergy of depth data and skeletons in combination with the strengths of graph based matching algorithms is an ongoing and future oriented research area. Hence, there is an inexhaustible potential of further scientific and technical subjects which are going to be investigated in the future.

Abbreviations

Symbols

a Arbitrary identifier with context-sensitive meaning. 14, 21–23, 25, 40, 53–59, 62, 83, 99, 111, 112, 135, 143, 144, 151, 188, 189, 194, 195

$\mathcal{A}$ Entity identifier (e.g. region, curve, configuration, etc.). 104, 105, 149, 150, 152

A Identifier with context-sensitive meaning. 18–20, 58, 59, 83, 84, 166

α Angle value. 39, 40, 95, 103–105, 116–118, 149

$\mathbf{A}$ Matrix identifier with context-sensitive meaning or affinity matrix. 21, 22, 51, 52, 121, 151, 152

$\mathbf{a}$ Arbitrary vector with context-sensitive meaning. 21, 22, 111, 112, 133, 194, 195

β Angle value. 116–118, 149

b Identifier or vector element of $\mathbf{b}$. 21–23, 25, 40, 53–59, 99, 111, 112, 143, 144, 151

$\mathcal{B}$ Entity identifier (e.g. region, curve, configuration, etc.). 104, 105

B Identifier with context-sensitive meaning. 18, 19, 42, 58, 59, 83, 84

$\mathbf{b}$ Point or rather vector in 2D and 3D. 21, 22

$c(\cdot)$ Cost function. 58, 59, 66, 67, 93–95, 118, 145, 151

$\mathbf{C}$ Cost matrix with context-sensitive meaning. 17, 18, 50, 53, 54, 56, 57, 67, 68, 70, 93, 94, 109

C Curve in 2D and 3D. 39, 40, 93, 94, 102, 127–129, 145, 147

δ Angle value 149

$\mathcal{D}$ Entity identifier (e.g. region, curve, configuration, etc.). 25, 26, 29, 44, 51, 139

$\delta(\cdot)$ Dirac function. 136

$d(\cdot)$ Distance function. 25, 26, 53, 56–59, 66, 67, 150, 151

D Distance Transform. 26, 40, 65, 128, 129, 156, 158

$\mathbb{D}$ Arbitrary set identifier with context-sensitive meaning. 1, 88, 98, 99, 105–107, 121, 134, 138, 160

e Identifier or vector element of **e**. 18, 23, 25, 50, 166

$\mathcal{E}$ Energy function. 135

ϵ Identifier with context-sensitive meaning. 78, 79, 81, 82, 158, 159, 177

η Identifier with context-sensitive meaning. 66, 135, 136, 147, 172, 189

e   Point or rather vector in 2D and 3D. 79, 80, 83–86, 107, 116–118, 149

$\mathcal{F}$ Feature set or space. 102, 156, 158, 160–162

f Function with contextual meaning. 27, 52, 56, 57, 79, 80, 83, 103, 106, 107, 121, 152, 165

F   Flow matrix of the Earth Mover's Distance. 58, 59

$F(\cdot)$ Fourier Transform. 80

γ Angle value. 52, 149

$g(\cdot)$ Function with contextual meaning. 21, 51, 52, 62, 83, 107, 109, 135, 136, 149, 160, 188

G Denotes an arbitrary contextual graph structure. 18, 22, 23, 25, 51, 56, 57, 64, 66, 68, 73, 75, 92, 118, 121, 122, 149–152, 199, 200

$H(\cdot)$ Heaviside function. 135, 136

h Arbitrary identifier with context-sensitive meaning. 138, 140–142, 160

$\mathcal{H}$ Hilbert Space 139, 140

H   Histogram 88, 89, 93

I   Image identifier. Qualifies both colour images and depth maps 100, 102, 135, 136, 188

$\hat{k}(\cdot)$ Kernel function. 139–141

k Identifier with context-sensitive meaning. 20, 29, 30, 93, 94, 131, 134, 137, 138, 140, 141, 194

$\mathcal{K}$ Entity identifier (e.g. region, curve, configuration, etc.). 23, 29, 44, 51, 83, 100, 101, 139, 147, 194, 195

K Length or size indicator of a set, vector or matrix dimension. 53, 54, 58, 59, 64–67, 73, 92–94, 98, 102, 116, 121, 149, 164

$l(\cdot)$ Function that returns the length of a path. 62, 66, 165

ϕ level set function. 127–132, 135, 136

$\mathcal{M}$ Matching identifier used for qualifying sets of correspondences. 19, 50

m Identifier with context-sensitive meaning. 29, 30, 41–43, 65, 94, 152

M Length or size indicator of a set, vector or matrix dimension. 53, 54, 62, 89, 93, 94, 116, 117

$\mathbf{M}$ Mutual Exclusion Constraints Matrix required in context of Maximum Weight Cliques 51, 52, 152

μ Threshold value with context-sensitive meaning. 82, 83, 106, 109–112, 135, 136, 147, 182, 186–188

N Length or size indicator of a set, vector or matrix dimension. 17, 18, 23, 30, 53, 54, 56–59, 65–67, 73, 79, 80, 89, 92–94, 99, 114, 121, 134, 138, 140, 141, 145, 147, 149, 164, 166, 177

$\mathbf{n}$ Normal vector. 39, 40, 129, 130, 149, 152

Ω Domain or object in 2D and 3D 29, 39–42, 62, 65, 66, 73, 80, 83, 88, 92, 93, 95, 127–129, 135, 143, 144, 156, 158, 160, 192–195, 197

p Identifier or vector element of $\mathbf{p}$. 1, 17, 18, 30, 40, 99, 106, 107, 119, 122, 161, 165

$\mathbf{P}$ Matrix identifier with contextual meaning. 114, 115

$\mathcal{P}^{[2,3]}$ Point cloud in 2D and 3D. 100–107, 109, 110

$\mathcal{P}$ 2D polygon. 62, 63, 81, 82

P Precision value of a retrieval system. 29, 30

ψ Identifier with context-sensitive meaning. 130, 138, 140, 141

$\psi(\cdot)$ Function with contextual meaning. 23, 25, 50, 54, 62, 107, 132, 150, 151, 156, 157, 160, 182, 194, 195

$\mathbf{p}$ Point or rather vector in 2D and 3D. 25, 26, 40, 44, 62, 63, 65, 78, 79, 81, 89, 92, 93, 105–107, 111, 112, 119–121, 142, 144–147, 149, 156, 158–162, 166

q Identifier or vector element of $\mathbf{p}$. 1, 17, 18, 106, 164, 165

Q Entity identifier (e.g. region, curve, configuration, etc.). 134, 138

$\mathbf{q}$ Point or rather vector in 2D and 3D. 25, 26, 65, 81–83, 105–107, 111, 112, 134, 138, 140, 149, 156, 164, 165

R Recall value of a retrieval system. 29, 30

r Identifier with context-sensitive meaning. 30, 40, 53, 54, 65, 66, 121, 122, 152

ρ Shortest path between two points. 22, 23, 64, 66, 67, 149–151, 160, 165, 167

Γ Set of shortest paths. 160

S Skeleton in 2D and 3D. 40, 42, 48, 66, 158, 160, 162, 165, 166

τ Identifier with context-sensitive meaning. 48, 135, 136, 140, 141, 145–147, 189

θ Parameter value or vector with context-sensitive meaning. 98, 100, 101, 121, 152, 190

T Time-crossing map. 27, 82–84, 133, 143–148, 189

t Depicts a single time step of a context-sensitive evolution process. 14, 39, 40, 52, 62, 63, 82–84, 98, 107, 128, 130–132, 135, 136, 138, 140, 141, 145, 147

u Identifier with context-sensitive meaning. 51, 66, 67, 73, 114, 118, 121, 122, 145, 149–152, 200

υ Velocity function or field. 27, 128, 130, 131, 133, 143–145

v Identifier or vector element of $\mathbf{v}$. 18–20, 25, 51, 64, 66, 67, 73, 75, 118, 119, 121, 122, 149–152, 200

List of Figures

List of Tables

Bibliography

[AT05]   Cagri Aslan and Sibel Tari. "An Axis-Based Representation for Recognition". In: *Proceedings of the Tenth IEEE International Conference on Computer Vision.* 2005, pp. 1339–1346.

[Abr+09]   Iosief Abraha, Carlo Romagnoli, Alessandro Montedori, and Roberto Cirocchi. "Thoracic stent graft versus surgery for thoracic aneurysm". In: *Cochrane Database of Systematic Reviews* 1 (2009).

[Au+08]   Oscar Kin-Chung Au, Chiew-Lan Tai, Hung-Kuo Chu, Daniel Cohen-Or, and Tong-Yee Lee. "Skeleton Extraction by Mesh Contraction". In: *ACM SIGGRAPH 2008 Papers.* SIGGRAPH '08. ACM, 2008, 44:1–44:10.

[BET09]   E. Baseski, A. Erdem, and S. Tari. "Dissimilarity between two skeletal trees in a context". In: *Pattern Recognition* 42.3 (2009), pp. 370–385.

[BF81]   Robert C. Bolles and Martin A. Fischler. "A RANSAC-Based Approach to Model Fitting and Its Application to Finding Cylinders in Range Data". In: *Proceedings of the 7th International Joint Conference on Artificial Intelligence (IJCAI '81), Vancouver, BC, Canada, August 1981.* 1981, pp. 637–643.

[BHR00]   Lasse Bergroth, Harri Hakonen, and Timo Raita. "A Survey of Longest Common Subsequence Algorithms". In: *String Processing and Information Retrieval, 2000. SPIRE 2000. Proceedings. 7th International Symposium on.* 2000, pp. 39–48.

[BI04]   Angelika Brennecke and Tobias Isenberg. "3D Shape Matching Using Skeleton Graphs". In: *Simulation and Visualization.* 2004, pp. 299–310.

[BL08]   Xiang Bai and Longin J. Latecki. "Path Similarity Skeleton Graph Matching". In: *IEEE Transactions on Pattern Analysis and Machine Intelligence* 30.7 (2008), pp. 1282–1292.

[BLL07]   Xiang Bai, Longin J. Latecki, and Wen-yu Liu. "Skeleton Pruning by Contour Partitioning with Discrete Curve Evolution". In: *IEEE Transactions on Pattern Analysis and Machine Intelligence* 29.3 (2007), pp. 449–462.

[BLT09]   Xiang Bai, Wenyu Liu, and Zhuowen Tu. "Integrating contour and skeleton for shape classification". In: *Computer Vision Workshops (ICCV Workshops), 2009 IEEE 12th International Conference on.* 2009, pp. 360–367.

[BMP02]   Serge Belongie, Jitendra Malik, and Jan Puzicha. "Shape matching and object recognition using shape contexts". In: *Pattern Analysis and Machine Intelligence, IEEE Transactions on* 24.4 (2002), pp. 509–522.

[BN10]   Prabin Bariya and Ko Nishino. "Scale-Hierarchical 3D Object Recognition in
            Cluttered Scenes". In: *Computer Vision and Pattern Recognition (CVPR), 2010
            IEEE Conference on.* 2010, pp. 1657–1664.

[BRF]   Liefeng Bo, Xiaofeng Ren, and Dieter Fox. "Depth kernel descriptors for
            object recognition". In: *Robotics and Automation (ICRA), 2011 IEEE International
            Conference on.* IEEE, pp. 821–826.

[BRF12]   Liefeng Bo, Xiaofeng Ren, and Dieter Fox. "Unsupervised Feature Learn-
            ing for RGB-D Based Object Recognition". In: *In International Symposium on
            Experimental Robotics (ISER).* Springer, 2012, pp. 387–402.

[Bai+09]   Xiang Bai, Xinggang Wang, Longin J. Latecki, Wenyu Liu, and Zhuowen Tu.
            "Active skeleton for non-rigid object detection". In: *Computer Vision, 2009
            IEEE 12th International Conference on.* 2009, pp. 575–582.

[Bal12]   Guillaume Bal. *Introduction to Inverse Problems.* Tech. rep. Introduction to In-
            verse Problems, 2012.

[Bel54]   Richard Bellman. "The theory of dynamic programming". In: *Bull. Amer.
            Math. Soc.* 60.6 (1954), pp. 503–515.

[Blu+12]   M. Blum, J.T. Springenberg, J. Wulfing, and M. Riedmiller. "A Learned Feature
            Descriptor for Object Recognition in RGB-D Data". In: *Robotics and Automation
            (ICRA), 2012 IEEE International Conference on.* 2012, pp. 1298–1303.

[Blu67a]   Harry Blum. "A Transformation for Extracting New Descriptors of Shape".
            In: *Models for the Perception of Speech and Visual Form.* Cambridge: MIT Press,
            1967, pp. 362–380.

[Blu67b]   Harry Blum. "A transformation for extracting new descriptors of shape". In:
            ed. by W. Wathen-Dunn. Vol. Models for the Perception of Speech and Visual
            For. MIT Press, 1967, pp. 362–380.

[Blu73]   Harry Blum. "Biological shape and visual science (part I)". In: *Journal of
            Theoretical Biology* 38.2 (1973), pp. 205–287.

[Bo+11]   Liefeng Bo, Kevin Lai, Xiaofeng Ren, and Dieter Fox. "Object recognition with
            hierarchical kernel descriptors". In: *Computer Vision and Pattern Recognition
            (CVPR), 2011 IEEE Conference on.* 2011, pp. 1729–1736.

[Bor96]   Gunilla Borgefors. "On Digital Distance Transforms in Three Dimensions".
            In: *Computer Vision and Image Understanding* 64.3 (1996), pp. 368–376.

[Bur+98]   Rainer E. Burkard, Eranda Cela, Panos M. Pardalos, and Leonidas S. Pitsoulis.
            "The Quadratic Assignment Problem". In: *Handbook of Combinatorial Optimiza-
            tion.* Ed. by Panos M. Pardalos and Ding-Zhu (Eds.) Du. Kluwer Academic
            Publishers, 1998, pp. 241–338.

[Bær01]   Jakob Andreas Bærentzen. *On the Implementation of Fast Marching Methods
            for 3D Lattices.* Tech. rep. Technical University of Denmark - Department of
            Mathematical Modelling, 2001.

[CB07]   H. Chen and B. Bhanu. "3D Free-form Object Recognition in Range Im-
            ages Using Local Surface Patches". In: *Pattern Recognition Letters* 28.10 (2007),
            pp. 1252–1262.

[CF01]   Richard J. Campbell and Patrick J. Flynn. "A Survey of Free-form Object
            Representation and Recognition Techniques". In: *Computer Vision and Image
            Understanding* 81.2 (2001), pp. 166–210.

[CKS97]   Vicent Caselles, Ron Kimmel, and Guillermo Sapiro. "Geodesic Active Con-
            tours". In: *Int. J. Comput. Vision* 22.1 (1997), pp. 61–79.

[CNT13]   Yi Chen, Nasser M. Nasrabadi, and Trac D. Tran. "Hyperspectral Image Clas-
            sification via Kernel Sparse Representation". In: *Geoscience and Remote Sensing,
            IEEE Transactions on* 51.1 (2013), pp. 217–231.

[CSM07]   Nicu D. Cornea, Deborah Silver, and Patrick Min. "Curve-Skeleton Proper-
            ties, Applications, and Algorithms". In: *IEEE Transactions on Visualization and
            Computer Graphics* 13.3 (2007), pp. 530–548.

[Cao+10]   Junjie Cao, Andrea Tagliasacchi, Matt Olson, Hao Zhang, and Zhinxun Su.
            "Point Cloud Skeletons via Laplacian Based Contraction". In: *Shape Modeling
            International Conference (SMI), 2010*. 2010, pp. 187–197.

[Cao+11]   Yu Cao, Zhiqi Zhang, Irina Czogiel, Ian Dryden, and Song Wang. "2D nonrigid
            partial shape matching using MCMC and contour subdivision". In: *Computer
            Vision and Pattern Recognition, IEEE Conference on*. 2011, pp. 2345–2352.

[Cor+05a]   Nicu D. Cornea, M.F. Demirci, Deborah Silver, Ali Shokoufandeh, Sven J.
            Dickinson, and Paul B. Kantor. "3D Object Retrieval using Many-to-many
            Matching of Curve Skeletons". In: *Shape Modeling and Applications, 2005 Inter-
            national Conference*. 2005, pp. 366–371.

[Cor+05b]   Nicu D. Cornea, Deborah Silver, Xiaosong Yuan, and Raman Balasubrama-
            nian. "Computing hierarchical curve-skeletons of 3D objects". In: *The Visual
            Computer* 21 (11 2005), pp. 945–955.

[DP73]   David H. Douglas and Thomas K. Peucker. "Algorithms for the Reduction of
            the Number of Points Required to Represent a Digitized Line or its Carica-
            ture". In: *Cartographica: The International Journal for Geographic Information and
            Geovisualization* 10.2 (1973), pp. 112–122.

[DSD09]   M. Fatih Demirci, Ali Shokoufandehand, and Sven Dickinson. "Skeletal Shape
            Abstraction from Examples". In: *Pattern Analysis and Machine Intelligence, IEEE
            Transactions on* 31.5 (2009), pp. 944–952.

[DT97]   George B. Dantzig and Mukund N. Thapa. *Linear Programming 1: Introduction*.
            Secaucus, NJ, USA: Springer-Verlag New York, Inc., 1997. ISBN: 0-387-94833-3.

[Dem+06]   M. Fatih Demirci, Ali Shokoufandeh, Yakov Keselman, Lars Bretzner, and
            Sven J. Dickinson. "Object Recognition as Many-to-Many Feature Matching".
            In: *International Journal of Computer Vision* 69 (2006), pp. 203–222.

[Dij59]   Edsger W. Dijkstra. "A note on two problems in connexion with graphs". In:
            Numerische Mathematik 1 (1 1959), pp. 269–271.

[Don+03]   Klaus Donath, Matthias Wolf, Radim Chrástek, and Heinrich Niemann. "A
            Hybrid Distance Map Based and Morphologic Thinning Algorithm". In: *Pat-
            tern Recognition*. Ed. by Bernd Michaelis and Gerald Krell. Vol. 2781. Lecture
            Notes in Computer Science. Springer Berlin Heidelberg, 2003, pp. 354–361.

[Dro+10]   Bertram Drost, Markus Ulrich, Nassir Navab, and Slobodan Ilic. "Model
            Globally, Match Locally: Efficient and Robust 3D Object Recognition". In:
            Computer Vision and Pattern Recognition (CVPR), 2010 IEEE Conference on. 2010,
            pp. 998–1005.

[Du+12]   Liang Du, Bo Ding, Shun Miao, Marcus Pfister, and Rui Liao. "Visual check
            and automatic compensation for patient movement during image-guided
            Abdominal Aortic Aneurysm (AAA) stenting". In: *Biomedical Engineering and
            Informatics (BMEI), 2012 5th International Conference on*. 2012, pp. 391–394.

[Eve+10]   Mark Everingham, Luc Van Gool, Christopher K.I. Williams, John Winn, and
            Andrew Zisserman. "The Pascal Visual Object Classes (VOC) Challenge". In:
            International Journal of Computer Vision 88.2 (2010), pp. 303–338.

[FJS10]   Vittorio Ferrari, Frederic Jurie, and Cordelia Schmid. "From Images to Shape
            Models for Object Detection". In: *International Journal of Computer Vision* 87.3
            (2010), pp. 284–303.

[FTVG04]   Victorio Ferrari, Tinne Tuytelaars, and Luc J. Van Gool. "Integrating multiple
            model views for object recognition". In: *Computer Vision and Pattern Recogni-
            tion, 2004. CVPR 2004. Proceedings of the 2004 IEEE Computer Society Conference
            on*. Vol. 2. 2004, II–105–II–112 Vol.2.

[Fab+08a]   Ricardo Fabbri, Luciano Da F. Costa, Julio C. Torelli, and Odemir M. Bruno.
            "2D Euclidean Distance Transform Algorithms: A Comparative Survey". In:
            ACM Comput. Surv. 40.1 (2008), 2:1–2:44.

[Fab+08b]   Ricardo Fabbri, Luciano Da F. Costa, Julio C. Torelli, and Odemir M. Bruno.
            "2D Euclidean distance transform algorithms: A comparative survey". In:
            ACM Computing Surveys 40.1 (2008), pp. 1–44.

[Fel+10]   P. F. Felzenszwalb, R. B. Girshick, D. McAllester, and D. Ramanan. "Ob-
            ject Detection with Discriminatively Trained Part-Based Models". In: *Pattern
            Analysis and Machine Intelligence, IEEE* 32.9 (2010), pp. 1627–1645.

[Fes09]   Paola Festa. *The shortest path tour problem: problem definition, modeling, and opti-
            mization*. Tech. rep. Department of Mathematics and Applications, University
            of Napoli FEDERICO II, 2009.

[Fri+00]   Sarah F. Frisken, Ronald N. Perry, Alyn P. Rockwood, and Thouis R. Jones.
            "Adaptively Sampled Distance Fields: A General Representation of Shape for
            Computer Graphics". In: *Proceedings of the 27th Annual Conference on Computer
            Graphics and Interactive Techniques*. SIGGRAPH. 2000, pp. 249–254.

[GK09]   Faming Gong and Cui Kang. "3D Mesh Skeleton Extraction Based on Feature
            Points". In: *Computer Engineering and Technology, 2009. ICCET '09. International
            Conference on*. Vol. 1. 2009, pp. 326–329.

[GW06]   Rafael C. Gonzalez and Richard E. Woods. *Digital Image Processing (3rd Edi-
            tion)*. 3rd ed. Upper Saddle River, NJ, USA: Prentice-Hall, Inc., 2006. ISBN:
            013168728X.

[Gal07]   David Gale. "Linear programming and the simplex method". In: *Notices of the
            AMS*. Vol. 54. 3. 2007, pp. 364–369.

[Gar+06]   Santiago. Garrido, Luis. Moreno, M. Abderrahim, and F. Martin. "Path Plan-
            ning for Mobile Robot Navigation using Voronoi Diagram and Fast March-
            ing". In: *Intelligent Robots and Systems, 2006 IEEE/RSJ International Conference
            on*. 2006, pp. 2376–2381.

[HBK01]   Masayuki Hisada, Alexander G. Belyaev, and Tosiyasu L. Kunii. "A 3D
            Voronoi-based skeleton and associated surface features". In: *Computer Graph-
            ics and Applications. Proceedings. Ninth Pacific Conference on*. 2001, pp. 89–96.

[HF05]   M.S. Hassouna and Aly A. Farag. "Robust centerline extraction framework
            using level sets". In: *Computer Vision and Pattern Recognition, 2005. CVPR 2005.
            IEEE Conference on*. Vol. 1. 2005, pp. 458–465.

[HF07a]   M. Sabry Hassouna and Aly A. Farag. "MultiStencils Fast Marching Methods:
            A Highly Accurate Solution to the Eikonal Equation on Cartesian Domains."
            In: *IEEE Trans. Pattern Anal. Mach. Intell.* 29.9 (2007), pp. 1563–1574.

[HF07b]   M.S. Hassouna and Aly A. Farag. "On the Extraction of Curve Skeletons
            using Gradient Vector Flow". In: *Computer Vision, 2007. ICCV 2007. IEEE 11th
            International Conference on*. 2007, pp. 1–8.

[HMS08]   Daniel Hartmann, Matthias Meinke, and Wolfgang Schröder. "Differential
            equation based constrained reinitialization for level set methods". In: *Journal
            of Computational Physics* 227.14 (2008), pp. 6821–6845.

[HZ03]   Richard Hartley and Andrew Zisserman. *Multiple View Geometry in Computer
            Vision*. 2nd ed. New York, NY, USA: Cambridge University Press, 2003. ISBN:
            0521540518.

[Han+13]   Jungong Han, Ling Shao, Dong Xu, and J. Shotton. "Enhanced Computer
            Vision With Microsoft Kinect Sensor: A Review". In: *Cybernetics, IEEE Trans-
            actions on* 43.5 (2013), pp. 1318–1334.

[Hay+11]   Tomoki Hayashi, Benjamin Raynal, Vincent Nozick, and Hideo Saito. "Skele-
            ton Features Distribution for 3D Object Retrieval". In: *proc. of the 12th IAPR
            Machine Vision and Applications (MVA2011)*. 2011, pp. 377–380.

[JH99]   Andrew E. Johnson and Martial Hebert. "Using Spin Images for Efficient
            Object Recognition in Cluttered 3D Scenes". In: *Pattern Analysis and Machine
            Intelligence, IEEE Transactions on* 21.5 (1999), pp. 433–449.

[Jan+11]   A. Janoch, S. Karayev, Yangqing Jia, J.T. Barron, M. Fritz, K. Saenko, and T.
            Darrell. "A category-level 3-D object dataset: Putting the Kinect to work". In:
            *Computer Vision Workshops (ICCV Workshops), 2011 IEEE International Confer-
            ence on*. 2011, pp. 1168–1174.

[Joh11]   Niklas Johansson. *Implementation of a standard level set method for incompress-
            ible two-phase flow simulations*. Tech. rep. Disciplinary Domain of Science and
            Technology, Uppsala University, 2011.

[KDB10]   Peter Kontschieder, Michael Donoser, and Horst Bischof. "Beyond Pairwise
            Shape Similarity Analysis". English. In: *Computer Vision - ACCV 2009*. Ed. by
            Hongbin Zha, Rin-ichiro Taniguchi, and Stephen Maybank. Vol. 5996. Lecture
            Notes in Computer Science. Springer Berlin Heidelberg, 2010, pp. 655–666.

[KGV83]   S. Kirkpatrick, C. D. Gelatt, and M. P. Vecchi. "Optimization By Simulated Annealing". In: *Science* 220 (1983), pp. 671–680.

[KLT05]   Sagi Katz, George Leifman, and Ayellet Tal. "Mesh segmentation using feature point and core extraction". In: *The Visual Computer* 21.8–10 (2005), pp. 649–658.

[KN09]   Sven Oliver Krumke and Hartmut Noltemeier. *Graphentheoretische Konzepte und Algorithmen*. 2009.

[KP07]   Jacek Kawa and Ewa Pietka. "Kernelized Fuzzy c-means Method in Fast Segmentation of Demyelination Plaques in Multiple Sclerosis". In: *Engineering in Medicine and Biology Society, 2007. EMBS 2007. 29th Annual International Conference of the IEEE*. 2007, pp. 5616–5619.

[KQ04]   Cemil Kirbas and Francis Quek. "A Review of Vessel Extraction Techniques and Algorithms". In: *ACM Computing Surveys (CSUR)* 36.2 (2004), pp. 81–121.

[KSK01]   Philip N. Klein, Thomas B. Sebastian, and Benjamin B. Kimia. "Shape matching using edit-distance: an implementation". In: *Proceedings of the twelfth annual ACM-SIAM symposium on Discrete algorithms*. Philadelphia, PA, USA: Society for Industrial and Applied Mathematic, 2001, pp. 781–790.

[KWT88]   Michael Kass, A. Witkin, and Demetri Terzopoulos. "Snakes: Active contour models". In: *International Journal of Computer Vision* 1.4 (1988), pp. 321–331.

[Kle+00]   Philip Klein, Tirthapura Srikanta, Daniel Sharvit, and Ben Kimia. "A tree-edit-distance algorithm for comparing simple, closed shapes". In: *Proceedings of the eleventh annual ACM-SIAM symposium on Discrete algorithms*. Philadelphia, PA, USA: Society for Industrial and Applied Mathematic, 2000, pp. 696–704.

[Kuh55]   Harold W. Kuhn. "The Hungarian Method for the Assignment Problem". In: *Naval Research Logistics Quarterly* 2.1–2 (1955), pp. 83–97.

[LF09]   Kevin Lai and Dieter Fox. "3D laser scan classification using web data and domain adaptation." In: *Robotics: Science and Systems*. 2009.

[LFU13]   Dahua Lin, Sanja Fidler, and Raquel Urtasun. "Holistic Scene Understanding for 3D Object Detection with RGBD Cameras". In: *Computer Vision (ICCV), 2013 IEEE International Conference on*. 2013, pp. 1417–1424.

[LG07]   Xinju Li and Igor Guskov. "3D Object Recognition from Range Images Using Pyramid Matching". In: *Computer Vision, 2007. ICCV 2007. IEEE 11th International Conference on*. 2007, pp. 1–6.

[LL00]   Longin J. Latecki and Rolf Lakämper. "Shape Similarity Measure Based on Correspondence of Visual Parts". In: *IEEE Trans. Pattern Anal. Mach. Intell.* 22.10 (2000), pp. 1185–1190.

[LL99a]   Longin J. Latecki and Rolf Lakämper. "Convexity Rule for Shape Decomposition Based on Discrete Contour Evolution." In: *Computer Vision and Image Understanding* 73.3 (1999), pp. 441–454.

[LL99b]   Longin J. Latecki and Rolf Lakämper. "Polygon Evolution by Vertex Deletion". In: *Proceedings of the Second International Conference on Scale-Space Theories in Computer Vision*. SCALE-SPACE '99. Springer-Verlag, 1999, pp. 398–409. ISBN: 3-540-66498-X.

[LLS92]   Louisa Lam, Seong-Whan Lee, and Ching Y. Suen. "Thinning Methodologies
            - A Comprehensive Survey". In: *IEEE Trans. Pattern Anal. Mach. Intell.* 14.9
            (1992), pp. 869–885.

[LS10]   Joerg Liebelt and Cordelia Schmid. "Multi-view object class detection with a
            3D geometric model". In: *IEEE CVPR.* 2010, pp. 1688–1695.

[LS13]   Li Liu and Ling Shao. "Learning discriminative representations from RGB-D
            video data". In: *Proceedings of the Twenty-Third international joint conference on
            Artificial Intelligence.* AAAI Press. 2013, pp. 1493–1500.

[LZL04]   John Lafferty, Xiaojin Zhu, and Yan Liu. "Kernel Conditional Random Fields:
            Representation and Clique Selection". In: *Proceedings of the Twenty-first Inter-
            national Conference on Machine Learning.* ICML '04. ACM, 2004.

[Lai+11a]   Kevin Lai, Liefeng Bo, Xiaofeng Ren, and Dieter Fox. "A large-scale hierar-
            chical multi-view RGB-D object dataset". In: *Robotics and Automation (ICRA),
            2011 IEEE International Conference on.* IEEE, 2011, pp. 1817–1824.

[Lai+11b]   Kevin Lai, Liefeng Bo, Xiaofeng Ren, and Dieter Fox. "Sparse distance learn-
            ing for object recognition combining RGB and depth information". In: *Robotics
            and Automation (ICRA), IEEE International Conference on.* 2011, pp. 4007–4013.

[Lat+07a]   Longin J. Latecki, Vasileios Megalooikonomou, Qiang Wang, and Deguang
            Yu. "An Elastic Partial Shape Matching Technique". In: *Pattern Recognition*
            40.11 (2007), pp. 3069–3080.

[Lat+07b]   Longin J. Latecki, Qiang Wang, Suzan Koknar-Tezel, and Vasileios Mega-
            looikonomou. "Optimal Subsequence Bijection". In: *Data Mining, 2007. ICDM
            2007. Seventh IEEE International Conference on.* IEEE, 2007, pp. 565–570.

[Les+09]   David Lesage, Elsa D. Angelini, Isabelle Bloch, and Gareth Funka-Lea. "A
            review of 3D vessel lumen segmentation techniques: Models, features and
            extraction schemes". In: *Medical Image Analysis* 13.6 (2009), pp. 819 –845.

[Li+05]   Chunming Li, Chenyang Xu, Changfeng Gui, and Martin D. Fox. "Level
            set evolution without re-initialization: a new variational formulation". In:
            *Computer Vision and Pattern Recognition, 2005. CVPR 2005. IEEE Computer
            Society Conference on.* Vol. 1. 2005, pp. 430–436.

[Li+10]   Chunming Li, Chenyang Xu, Changfeng Gui, and Martin D. Fox. "Distance
            Regularized Level Set Evolution and Its Application to Image Segmentation".
            In: *Image Processing, IEEE Transactions on* 19.12 (2010), pp. 3243–3254.

[Lia+10]   Rui Liao, Yunhao Tan, Hari Sundar, Marcus Pfister, and Ali Kamen. "An Ef-
            ficient Graph-Based Deformable 2D/3D Registration Algorithm with Appli-
            cations for Abdominal Aortic Aneurysm Interventions". In: *Medical Imaging
            and Augmented Reality.* Vol. 6326. Lecture Notes in Computer Science. Springer
            Berlin Heidelberg, 2010, pp. 561–570.

[Llo82]   S. Lloyd. "Least squares quantization in PCM". In: *Information Theory, IEEE
            Transactions on* 28.2 (1982), pp. 129–137.

[Lu+10]   ChengEn Lu, Nagesh Adluru, Haibin Ling, Guangxi Zhu, and Longin J. Late-
            cki. "Contour Based Object Detection Using Part Bundles". In: *Computer Vision
            and Image Understanding* 114.7 (2010), pp. 827–834.

[MBH12]   W. Mohamed and A. Ben Hamza. "Reeb Graph Path Dissimilarity for 3D
            Object Matching and Retrieval". In: *The Visual Computer* 28.3 (2012), pp. 305–
            318.

[MBO06]   Ajmal S. Mian, Mohammed Bennamoun, and Robyn Owens. "Three-
            Dimensional Model-Based Object Recognition and Segmentation in Cluttered
            Scenes". In: *Pattern Analysis and Machine Intelligence, IEEE Transactions on* 28.10
            (2006), pp. 1584–1601.

[MJ13]   Kairanbay Magzhan and Hajar Mat Jani. "A Review And Evaluations Of
            Shortest Path Algorithms". In: *International Journal of Scientific and Technology
            Research* 2 (6 2013), pp. 99–104.

[ML11]   Tianyang Ma and Longin J. Latecki. "From partial shape matching through
            local deformation to robust global shape similarity for object detection". In:
            Computer Vision and Pattern Recognition (CVPR), 2011 IEEE Conference on. 2011,
            pp. 1441–1448.

[ML12]   Tianyang Ma and Longin J. Latecki. "Maximum Weight Cliques with Mutex
            Constraints for Video Object Segmentation". In: *CVPR*. 2012, pp. 670–677.

[MLP11]   Shun Miao, Rui Liao, and Marcus Pfister. "Toward smart utilization of two X-
            ray images for 2-D/3-D registration applied to abdominal aortic aneurysm
            interventions". In: *Biomedical Engineering and Informatics (BMEI), 2011 4th
            International Conference on*. Vol. 1. 2011, pp. 550–555.

[MS96]   Min C. Ma and Milan Sonka. "A Fully Parallel 3D Thinning Algorithm and
            Its Applications". In: *Computer Vision and Image Understanding, Vol. 64* (1996),
            pp. 420–433.

[MYL13]   Tianyang Ma, Meng Yi, and Longin J. Latecki. "View-Invariant Object Detec-
            tion by Matching 3D Contours". In: *ACCV Workshops*. Ed. by Jong-Il Park and
            Junmo Kim. Vol. 7729. Lecture Notes in Computer Science. Springer Berlin
            Heidelberg, 2013, pp. 183–196.

[Mac+02]   Diego Macrini, Ali Shokoufandeh, Sven Dickinson, Kaleem Siddiqi, and
            Steven Zucker. "View-Based 3-D Object Recognition using Shock Graphs". In:
            *Proceedings of the 16 th International Conference on Pattern Recognition (ICPR'02)
            Volume 3 - Volume 3*. ICPR '02. IEEE Computer Society, 2002, pp. 24–28.

[Mac67]   James B. MacQueen. "Some Methods for Classification and Analysis of Multi-
            Variate Observations". In: *Proc. of the fifth Berkeley Symposium on Mathematical
            Statistics and Probability*. Vol. 1. 1967, pp. 281–297.

[Mit02]   John E. Mitchell. "Branch-and-Cut Algorithms for Combinatorial Optimiza-
            tion Problems". In: *Handbook of Applied Optimization*. Oxford University Press,
            2002, pp. 65–77.

[Mül07]   Meinard Müller. *Information Retrieval for Music and Motion*. Secaucus, NJ, USA:
            Springer-Verlag New York, Inc., 2007. ISBN: 3540740473.

[NS09]   T. B. Nguyen and L. Sukhan. "Accurate 3D Lines Detection Using Stereo Camera". In: *Int. Sym. on Assembly and Manufacturing*. 2009, pp. 304–309.

[OI92]   Robert L. Ogniewicz and M. Ilg. "Voronoi skeletons: theory and applications". In: *Computer Vision and Pattern Recognition, 1992. Proceedings CVPR '92., IEEE Conference on*. 1992, pp. 63–69.

[OK95]   Robert L. Ogniewicz and Olaf Kübler. "Hierarchic Voronoi skeletons". In: *Pattern Recognition* 28.3 (1995), pp. 343–359.

[OS88]   Stanley Osher and James A. Sethian. "Fronts Propagating with Curvature-dependent Speed: Algorithms Based on Hamilton-Jacobi Formulations". In: *Journal of Computational Physics* 79.1 (1988), pp. 12–49.

[Ogn94]   R. L. Ogniewicz. "Skeleton-space: a multiscale shape description combining region and boundary information". In: *Computer Vision and Pattern Recognition, 1994. Proceedings CVPR '94., 1994 IEEE Computer Society Conference on*. 1994, pp. 746–751.

[PB07]   Norbert Pfeifer and Christian Briese. *Laser Scanning - Principles and Applications*. 2007.

[PK98]   K. Palágyi and A. Kuba. "A 3D 6-Subiteration Thinning Algorithm for Extracting Medial Lines". In: *Pattern Recognition Letters 19* (1998), pp. 613–627.

[PK99]   Kálmán Palágyi and Attila Kuba. "Directional 3D Thinning Using 8 Subiterations". In: *Proceedings of the 8th International Conference on Discrete Geometry for Computer Imagery*. DCGI '99. Springer-Verlag, 1999, pp. 325–336.

[POB87]   Stephen M. Pizer, William R. Oliver, and Sandra H. Bloomberg. "Hierarchical Shape Description Via the Multiresolution Symmetric Axis Transform". In: *Pattern Analysis and Machine Intelligence, IEEE Transactions on* 9.4 (1987), pp. 505–511.

[PSM94]   Pietro Perona, Takahiro Shiota, and Jitendra Malik. "Anisotropic Diffusion". In: *Geometry-Driven Diffusion in Computer Vision*. Ed. by Bart M. ter Haar Romeny. Vol. 1. Computational Imaging and Vision. Springer Netherlands, 1994, pp. 73–92.

[PT11]   Nadia Payet and Sinisa Todorovic. "From Contours to 3D Object Detection and Pose Estimation". In: *Computer Vision (ICCV), 2011 IEEE International Conference on*. 2011, pp. 983–990.

[Pat10]   S. Patty. *Finite Difference Methods and Solving the Level Set Equations Numerically*. Tech. rep. Department of Mathematics, Brigham Young University, 2010.

[Pen07]   David W. Pentico. "Assignment problems: A golden anniversary survey". In: *European Journal of Operational Research* 176.2 (2007), pp. 774–793.

[Piz08]   Zygmunt Pizlo. *3D Shape: Its Unique Place in Visual Perception*. Cambridge - Massachusetts, London - England: The MIT Press, 2008.

[Qia+04]   Gang Qian, Shamik Sural, Yuelong Gu, and Sakti Pramanik. "Similarity Between Euclidean and Cosine Angle Distance for Nearest Neighbor Queries". In: *Proceedings of the 2004 ACM Symposium on Applied Computing*. SAC '04. New York, NY, USA: ACM, 2004, pp. 1232–1237.

[RC+12]   Carolina Redondo-Cabrera, Roberto J. Lopez-Sastre, Javier Acevedo-Rodriguez, and Saturnino Maldonado-Bascon. "SURFing the Point Clouds: Selective 3D spatial pyramids for Category-Level Object Recognition". In: *Computer Vision and Pattern Recognition (CVPR), 2012 IEEE Conference on*. 2012, pp. 3458–3465.

[RN09]   Stuart Russell and Peter Norvig. *Artificial Intelligence: A Modern Approach*. 3rd. Prentice Hall Press, 2009.

[RT06]   Dennie Reniers and Alexandru Telea. "Quantitative Comparison of Tolerance-Based Feature Transforms". In: *First International Conference on Computer Vision Theory and Applications (VISAPP)* (2006), pp. 107–114.

[RTG00]   Yossi Rubner, Carlo Tomasi, and Leonidas J. Guibas. "The Earth Mover's Distance As a Metric for Image Retrieval". In: *Int. J. Comput. Vision* 40.2 (2000), pp. 99–121. ISSN: 0920-5691.

[Rah+10]   Ali Raheem, Tom Carrell, Bijan Modarai, and Graeme Penney. "Non-rigid 2D-3D image registration for use in endovascular repair of abdominal aortic aneurysms". In: *Medical Image Understanding and Analysis*. 2010, pp. 153–157.

[Ren09]   Dennie Reniers. "Skeletonization and Segmentation of Binary Voxel Shapes". PhD thesis. Technische Universiteit Eindhoven, 2009.

[SA12]   Luciano Spinello and Kai O. Arras. "Leveraging RGB-D Data: Adaptive fusion and domain adaptation for object detection". In: *Robotics and Automation (ICRA), 2012 IEEE International Conference on*. 2012, pp. 4469–4474.

[SB98]   Doron Shaked and Alfred M. Bruckstein. "Pruning medial axes". In: *Comput. Vis. Image Underst.* 69.2 (1998), pp. 156–169.

[SBC05]   Jamie Shotton, Andrew Blake, and Roberto Cipolla. "Contour-based learning for object detection". In: *Computer Vision, 2005. ICCV 2005. Tenth IEEE International Conference on*. Vol. 1. 2005, pp. 503–510.

[SK05]   Thomas B. Sebastian and Benjamin B. Kimia. "Curves vs. skeletons in object recognition". In: *Signal Processing* 85.2 (2005), pp. 247–263.

[SK10]   Alexander Sabov and Jörg Krüger. "Identification and Correction of Flying Pixels in Range Camera Data". In: *Proceedings of the 24th Spring Conference on Computer Graphics*. SCCG '08. New York, NY, USA: ACM, 2010, pp. 135–142.

[SK96]   Kaleem Siddiqi and Benjamin B. Kimia. "A shock grammar for recognition". In: (1996), pp. 507–513.

[SKK01]   Thomas B. Sebastian, Philip N. Klein, and Benjamin B. Kimia. "Recognition of Shapes by Editing Shock Graphs". In: *In IEEE International Conference on Computer Vision*. 2001, pp. 755–762.

[SKK04]   Thomas B. Sebastian, Philip N. Klein, and Benjamin B. Kimia. "Recognition of Shapes by Editing Their Shock Graphs". In: *IEEE Transactions on Pattern Analysis and Machine Intelligence* 26.5 (2004), pp. 550–571.

[SP08]   Kaleem Siddiqi and Stephen Pizer. *Medial Representations: Mathematics, Algorithms and Applications*. 1st ed. Springer Publishing Company, Incorporated, 2008. ISBN: 1402086571, 9781402086571.

[SPB04]   Joaquim Salvi, Jordi Pagès, and Joan Batlle. "Pattern Codification Strategies in Structured Light Systems". In: *Pattern Recognition* 37.4 (2004), pp. 827 –849.

[SS10]   Shishir K. Shandilya and Nidhi Singhai. "A Survey On: Content Based Image Retrieval Systems". In: *International Journal of Computer Applications* 4.2 (2010), pp. 22–26.

[SSO94]   Mark Sussman, Peter Smereka, and Stanley Osher. "A Level Set Approach for Computing Solutions to Incompressible Two-Phase Flow". In: *Journal of Computational Physics* 114.1 (1994), pp. 146–159.

[Sch05]   Ingo Schmitt. *Ähnlichkeitssuche in Multimedia-Datenbanken - Retrieval, Suchalgorithmen und Anfragebehandlung.* Oldenbourg, 2005. ISBN: 3-486-57907-X.

[Set95]   James A. Sethian. "A Fast Marching Level Set Method for Monotonically Advancing Fronts". In: *Proc. Nat. Acad. Sci.* 1995, pp. 1591–1595.

[Set99a]   James A. Sethian. "Fast Marching Methods". In: *SIAM Rev.* 41.2 (1999), pp. 199–235.

[Set99b]   James A. Sethian. *Level Set Methods and Fast Marching Methods: Evolving Interfaces in Computational Geometry, Fluid Mechanics, Computer Vision, and Materials Science.* 2nd ed. Cambridge University Press, 1999. ISBN: 0521645573.

[Sha+07]   Andrei Sharf, Thomas Lewiner, Ariel Shamir, and Leif Kobbelt. "On-the-fly Curve-skeleton Computation for 3D Shapes". In: *Computer Graphics Forum* 26.3 (2007), pp. 323–328.

[She+11]   Wei Shen, Xiang Bai, Rong Hu, Hongyuan Wang, and Longin J. Latecki. "Skeleton Growing and Pruning with Bending Potential Ratio". In: *Pattern Recognition* 44.2 (2011), pp. 196–209.

[Sps]   *Accurate Fast Marching MATLAB Library.* http : / / www . mathworks . com / matlabcentral / fileexchange / 24531 - accurate - fast - marching. Online: 19th August 2015.

[Sti+06]   Stefan Stiene, Kai Lingemann, Andreas Nuchter, and Joachim Hertzberg. "Contour-Based Object Detection in Range Images". In: *Int. Sym. on 3D Data Processing, Visualization, and Transmission.* 2006, pp. 168–175.

[Sun+03]   H. Sundar, D. Silver, N. Gagvani, and S. Dickinson. "Skeleton Based Shape Matching and Retrieval". In: *Proceedings of the Shape Modeling International 2003.* SMI '03. IEEE Computer Society, 2003, pp. 130–139.

[TH02]   Roger Tam and Wolfgang Heidrich. "Feature-Preserving Medial Axis Noise Removal". In: *Computer Vision.* Ed. by Anders Heyden, Gunnar Sparr, Mads Nielsen, and Peter Johansen. Vol. 2351. Lecture Notes in Computer Science. Springer Berlin Heidelberg, 2002, pp. 672–686.

[TK08]   Sergios Theodoridis and Konstantinos Koutroumbas. *Pattern Recognition.* 4th ed. Academic Press, 2008. ISBN: 1597492728, 9781597492720.

[TNOL13]   Duc Thanh Nguyen, Philip O. Ogunbona, and Wanqing Li. "A Novel Shape-based Non-redundant Local Binary Pattern Descriptor for Object Detection". In: *Pattern Recognition* 46.5 (2013), pp. 1485–1500.

[TV98]   Emanuele Trucco and Alessandro Verri. *Introductory Techniques for 3-D Computer Vision*. Upper Saddle River, NJ, USA: Prentice Hall PTR, 1998. ISBN: 0132611082.

[TVD08]   Julien Tierny, Jean-Philippe Vandeborre, and Mohamed Daoudi. "Fast and precise kinematic skeleton extraction of 3D dynamic meshes". In: *Pattern Recognition, 2008. ICPR 2008. 19th International Conference on*. 2008, pp. 1–4.

[TZCO09]   Andrea Tagliasacchi, Hao Zhang, and Daniel Cohen-Or. "Curve Skeleton Extraction from Incomplete Point Cloud". In: *ACM SIGGRAPH 2009 Papers*. SIGGRAPH '09. New York, NY, USA: ACM, 2009, 71:1–71:9.

[Tad07]   Ryszard Tadeusiewicz. "What Does It Means Automatic Understanding of the Images?" In: *Imaging Systems and Techniques, 2007. IST '07. IEEE International Workshop on*. 2007, pp. 1–3.

[Tan+13]   Shuai Tang, Xiaoyu Wang, Xutao Lv, Tony X. Han, James Keller, Zhihai He, Marjorie Skubic, and Shihong Lao. "Histogram of Oriented Normal Vectors for Object Recognition with a Depth Sensor". In: *Proceedings of the 11th Asian Conference on Computer Vision - Volume Part II*. ACCV'12. Berlin, Heidelberg: Springer-Verlag, 2013, pp. 525–538.

[Tur09]   Volker Turau. *Algorithmische Graphentheorie (3. Auflage)*. Oldenbourg, 2009. ISBN: 978-3-486-59057-9.

[VUB07]   Robert Van Uitert and Ingmar Bitter. "Subvoxel Precise Skeletons of Volumetric Data Based on Fast Marching Methods". In: *Medical Physics* 34.2 (2007), pp. 627–638.

[WF74]   Robert A. Wagner and Michael J. Fischer. "The String-to-String Correction Problem". In: *J. ACM* 21.1 (1974), pp. 168–173.

[WL08]   Yu-Shuen Wang and Tong-Yee Lee. "Curve-Skeleton Extraction Using Iterative Least Squares Optimization". In: *IEEE Transactions on Visualization and Computer Graphics* 14.4 (2008), pp. 926–936.

[XWB09]   Yao Xu, Liu Wenyu, and Xiang Bai. "Skeleton Graph Matching Based on Critical Points Using Path Similarity". In: *Proceedings of the 9th Asian Conference on Computer Vision - Volume Part III*. ACCV'09. 2009, pp. 456–465.

[YCS00]   Peter J. Yim, Peter L. Choyke, and Roland M. Summers. "Gray-scale skeletonization of small vessels in magnetic resonance angiography". In: *Medical Imaging, IEEE Transactions on* 19.6 (2000), pp. 568–576.

[YKS07]   Pingkun Yan, Saad M. Khan, and Mubarak Shah. "3D Model based Object Class Detection in An Arbitrary View". In: *Computer Vision, 2007. ICCV 2007. IEEE 11th International Conference on*. 2007, pp. 1–6.

[YLL12]   Xingwei Yang, Hairong Liu, and Longin J. Latecki. "Contour-based object detection as dominant set computation". In: *Pattern Recognition* 45.5 (2012), pp. 1927–1936.

[Yan+09]   Xiaojun Yang, Xiang Bai, Xingwei Yang, and Luan Zeng. "An Efficient Quick Algorithm for Computing Stable Skeletons". In: *Image and Signal Processing, 2009. CISP '09. 2nd International Congress on*. 2009, pp. 1–5.

[Yi+13]   Meng Yi, Yinfei Yang, Wenjing Qi, Yu Zhou, Yunfeng Li, Zygmunt Pizlo, and Longin J. Latecki. "Navigation toward Non-static Target Object Using Footprint Detection Based Tracking". In: *Proceedings of the 11th ACCV*. ACCV. Springer-Verlag, 2013, pp. 389–400.

[You98]   Laurent Younes. "Computable Elastic Distances between Shapes". In: *SIAM Journal on Applied Mathematics* 58.2 (1998), pp. 565–586.

[ZS14]   Abdel Nasser H. Zaied and Laila Abd El fatah Shawky. "A Survey of Quadratic Assignment Problems". In: *International Journal of Computer Applications* 101.6 (2014), pp. 28–36.

[Zen+08]   Jingting Zeng, Rolf Lakaemper, Xingwei Yang, and Xin Li. "2D Shape Decomposition Based on Combined Skeleton-Boundary Features". In: *Proceedings of the 4th International Symposium on Advances in Visual Computing, Part II*. ISVC '08. Springer-Verlag, 2008, pp. 682–691.

[Zha+05]   Juan Zhang, Kaleem Siddiqi, Diego Macrini, Ali Shokoufandeh, and Sven Dickinson. "Retrieving articulated 3-d models using medial surfaces and their graph spectra". In: *Proceedings of the 5th international conference on Energy Minimization Methods in Computer Vision and Pattern Recognition*. EMMCVPR'05. Springer-Verlag, 2005, pp. 285–300.

[Zha+08]   Yan Zhang, Bogdan J. Matuszewski, Lik-Kwan Shark, and Christopher J. Moore. "Medical Image Segmentation Using New Hybrid Level-Set Method". In: *BioMedical Visualization, 2008. MEDIVIS '08. Fifth International Conference*. 2008, pp. 71–76.

[Zhu+06]   F. Zhuge, G. D. Rubin, S. H. Sun, and S. Napel. "An abdominal aortic aneurysm segmentation method: Level set with region and statistical information". In: *Medical Physics* 33.5 (2006), pp. 1440–1453.

[Zhu04]   Mu Zhu. *Recall, Precision and Average Precision*. Tech. rep. Department of Statistics and Actuarial Science, University of Waterloo, 2004.

[Zul12]   Marco Zuliani. *RANSAC for Dummies*. http : / / aiweb . techfak . uni - bielefeld.de/content/bworld-robot-control-software/. 2012.

[dBr+02]   Marleen de Bruijne, Bram van Ginneken, W. J. Niessen, J. B. Antoine Maintz, and M. A. Viergever. "Active-shape-model-based segmentation of abdominal aortic aneurysms in CTA images". In: *Proc. SPIE* 4684 (2002), pp. 463–474.

Own Publications

[AFG13] Ali Amanpourgharaei, Christian Feinen, and Marcin Grzegorzek. "Graph-Based Shape Representation for Object Retrieval". In: *2nd International Conference on Pattern Recognition, Applications and Methods*. Springer, Berlin, Heidelberg, 2013, pp. 315–318.

[Bad+13] Julian Bader, Christian Feinen, Jens Hedrich, Rodrigo Do Carmo, and Philipp Scholl. *Proceedings of the Joint Workshop of the German Research Training Groups in Computer Science*. Pro Business Digital Printing Deutschland GmbH, 2013.

[Cza+14] Joanna Czajkowska, Christian Feinen, Marcin Grzegorzek, M. Raspe, and R. Wickenhoefer. "A New Aortic Aneurysm CT Series Registration Algorithm". In: *International Conference Information Technologies in Biomedicine*. 2014.

[Fei+11] Christian Feinen, Marcin Grzegorzek, Detlev Droege, and Dietrich Paulus. "A Generic Approach to the Texture Detection Problem in Digital Images". In: *7th International Conference on Computer Recognition Systems*. Springer, Berlin, Heidelberg, 2011, pp. 375–384.

[Fei+13] Christian Feinen, Marcin Grzegorzek, David Barnowsky, and Dietrich Paulus. "Robust 3D Object Skeletonisation for the Similarity Measure". In: *2nd International Conference on Pattern Recognition, Applications and Methods*. Springer, Berlin, Heidelberg, 2013, pp. 167–175.

[Fei+14a] Christian Feinen, Joanna Czajkowska, Marcin Grzegorzek, and Longin J. Latecki. "3D Object Retrieval by 3D Curve Matching". In: *IEEE International Conference on Image Processing*. IEEE Computer Society, 2014.

[Fei+14b] Christian Feinen, Joanna Czajkowska, Marcin Grzegorzek, and Longin J. Latecki. "Computer Vision and Machine Learning with RGB-D Sensors". In: Springer, 2014. Chap. Matching of 3D Objects Based on 3D Curves.

[Fei+14c] Christian Feinen, Cong Yang, Oliver Tiebe, and Marcin Grzegorzek. "Shape Matching Using Point Context and Contour Segments". In: *Asian Conference on Computer Vision*. Springer, 2014, Singapore.

[Fei+14d] Christian Feinen, Joanna Czajkowska, Marcin Grzegorzek, Matthias Raspe, and R. Wickenhoefer. "Skeleton-Based Abdominal Aorta Registration Technique". In: *International Conference of the IEEE Engineering in Medicine and Biology Society*. 2014.

[Hed+13]   Jens Hedrich, Cong Yang, Christian Feinen, Simone Schaefer, Dietrich Paulus, and Marcin Grzegorzek. "Extended Investigations on Skeleton Graph Matching for Object Recognition". In: *8th International Conference on Computer Recognition Systems*. Springer LNCS, 2013, pp. 371–381.

[Yan+14]   Cong Yang, Oliver Tiebe, Pit Pietsch, Christian Feinen, Udo Kelter, and Marcin Grzegorzek. "Shape-Based Object Retrieval by Contour Segment Matching". In: *IEEE International Conference on Image Processing*. IEEE, 2014.

[Yan+15a]   Cong Yang, Christian Feinen, Oliver Tiebe, Kimiaki Shirahama, and Marcin Grzegorzek. "Shape-based Object Matching Using Point Context". In: *International Conference on Multimedia Retrieval*. 2015, pp. 519–522.

[Yan+15b]   Cong Yang, Oliver Tiebe, Pit Pietsch, Christian Feinen, Udo Kelter, and Marcin Grzegorzek. "Shape-based Object Retrieval and Classification with Supervised Optimisation". In: *International Conference on Pattern Recognition Applications and Methods*. Springer, 2015, pp. 204–211.